I0034218

GINI

Capitalism, Cryptocurrencies & the Battle for Human Rights

FERRIS EANFAR

All book sales proceeds support the
nonprofit, nonpartisan Gini Foundation.

First edition published in 2018 by The Gini Foundation,
a nonprofit foundation.
The Gini Foundation
Website address: GiniFoundation.org

THE LIBRARY OF CONGRESS HAS CATALOGUED
THIS BOOK AS FOLLOWS:

Names: Eanfar, Ferris.
Title: Gini : capitalism, cryptocurrencies & the battle for human rights / Ferris Eanfar.
Description: New York, NY : The Gini Foundation, 2018.
Identifiers: LCCN 2018914917 | 9780999112144 (hardcover) | 9780999112137
(paperback) | 9780999112151 (ebook)
Subjects: Cryptocurrencies. | Blockchains (Databases) | Economic history—21 st
century. | Income distribution. | Corporate power. | Wealth—Moral and ethical aspects. |
Land value taxation.
Classification: LCC HG1710 .E26 2018 | DDC 332.4—dc23

Dedication

This book is dedicated to all the humans on Earth who are suffering from broken capitalism and broken democracy today.

You deserve better. Let's build a better future together.

Contents

Acknowledgments

I appreciate all the thoughtful and accomplished people in my life who have given me encouragement over the years, but a few human angels were exceptionally encouraging during the development of this book. Words are not enough to capture my appreciation for them, but I must start somewhere; so, these words are a down payment toward the debt of gratitude I owe them all. I thank . . .

Jeff Knowles, for his unassuming intelligence, attention to detail, serenity under pressure, and appreciation for the philosophical aspects of our lives and professional projects. I also appreciate him for sharing with me his decades of technical knowledge and experience in the payment processing industry. Nobody knows that industry and how to build financial services technologies better than him. It's difficult to imagine a human with more complementary professional strengths to balance my weaknesses, which is one of the reasons we work so well together. I am blessed to have Jeff as one of my dearest friends and professional partners.

Jen Bawden, for her unyielding encouragement and enthusiasm for life and everything we do together. She is truly a force of nature. Within seconds, anybody who meets her can see that she is likely not from Earth. I have known her for years and I still wonder if she might be an angel sent from the heavens, temporarily inhabiting a gorgeous human shell, precision-engineered to attract the most thoughtful and intelligent souls on Earth into her orbit, so that she can complete her mission to elevate humanity to her angelic level of consciousness. Imagine a creature with the DNA of Mother Teresa, Joan of Arc, Rita Hayworth, Jesus Christ, Reese Witherspoon, Catherine Deneuve, a tenacious pit bull, a glittering unicorn, and a bright rainbow hugging the horizon—that's Jen, the Universe's gift to humanity.

The rest of our Gini team, for the extraordinary talent and experience that they bring to the Gini Foundation every day. The Gini technology is on the cutting-edge of what is possible in cryptocurrency engineering and I'm deeply grateful for having such a world-class group of thoughtful and bright humans who really understand why Gini is important in the world today. Having a team with the skills, experience and discipline to transform ideas and architectural designs into a secure and scalable platform for humanity is one of the many reasons why Gini is different from other cryptocurrencies in nearly every conceivable way.

Preface

This Book Should Not Exist. The economic dysfunction that is killing capitalism around the world today should not exist. The geopolitical dysfunction that is killing democracy around the world today should not exist. Yet, here we are, living and suffering under the crushing weight of increasingly dysfunctional and corrupt institutions that are choking the life, liberty, and wealth out of billions of humans on Earth today. That's why this book *must* exist.

The Evolution of This Book. I've been thinking about many of the technical, socioeconomic, and geopolitical principles in this book since the 9/11/2001 terrorist attacks. The underlying causes of those attacks and the subsequent tornado of unconstitutional actions taken by senior U.S. politicians after declaring their *War on Terror* have given me a constant sense of urgency. As a result, I've written many nonpartisan articles and several books about how to eliminate (or at least dramatically reduce) systemic corruption in the U.S. political and economic systems. I also published the Global Governance Scorecard, which compares dozens of countries across many important technical, economic, quality-of-life, and quality-of-governance dimensions. All that work and experience has laid the foundations for this book today.

The Evolution of the Platform. One of the books I wrote was never published because, when I was writing it back in 2013, the ideas of blockchains, cryptocurrencies, and decentralized voting systems were too radical for most people to understand, much less embrace as serious alternatives to the dysfunctional status quo. As a result, I referred to the concepts in that book as "The Platform" at Eanfar.org and in my other writings. That book evolved into this book after the publication of my previous book, *Broken Capitalism: This Is How We Fix It.*

This Is the Time. There are many historical cases of ideas and products that were too far ahead of their time, which caused them to fail due to lack of mainstream adoption. The Gini Foundation team didn't want to be caught in that trap; so, we waited for better timing. The research and analysis for the *Broken Capitalism* book revealed that public awareness of blockchains and cryptocurrencies seemed to be broad enough for the general public to embrace the systems we have wanted to build since 2013. Now, the time is right to launch the Gini Foundation based on the principles, philosophies, and technologies revealed in this book, which has been nearly two decades in the making.

The Relationship Between AngelPay and Gini. The AngelPay Foundation and the Gini Foundation are both nonprofit organizations launched by the same core team, on which I have the privilege to serve. These two foundations are focused on separate but complementary missions.

- **AngelPay Foundation.** For years, AngelPay has been focused on "returning wealth and power to the creators of value" by giving small- and medium-sized companies access to merchant payment processing services at nonprofit rates, which are typically only available to much larger companies. This is one of the ways that AngelPay supports the small- to medium-sized business community to offset the growing tyranny of ecosystem-killing corporate goliaths.
- **Gini Foundation.** The nonprofit Gini Foundation is focused on building cryptographically secure, high-integrity systems that protect human rights and privacy within the context of an equitable and sustainable monetary system that optimizes the broad-based, wealth-generating capacity of real-world commerce. Based on our research, no other organization today has Gini's combination of real-world experience, technical knowledge and skills, human and financial resources, and properly aligned incentives to build a truly egalitarian, equitable, and sustainable socioeconomic ecosystem.

Focused on Solutions. This is a book about solutions. Some of the solutions are technology-based, while other solutions presented are based on clearly defined, nonpartisan economic policies and human rights that all humans should expect in any civilized, democratic society.

I Respect Your Time. This book is intended to be thought-provoking, engaging, and sometimes entertaining, but the primary purpose of this book is to inspire action. Learning about problems and solutions without learning how to effectively implement the solutions is a waste of time. This book respects your time by delivering practical knowledge, timeless wisdom, and specific tools that can make a meaningful positive impact in your life and community.

Navigational Guides

Our Purpose. This book explores many interconnected economic, geopolitical, social, and technological challenges facing humans around the world today. Given the complexity of some of the topics in this book, we often need to dive deep into the heart of various issues. Our purpose here together is to wrap our minds around the core social, geopolitical, economic, and technical principles that are causing the problems described in each chapter so that we can understand and embrace the corresponding solutions.

Our Mission: Build A Lifeboat for Humanity. Exploring the deepest recesses of the human condition, wrestling with timeless existential questions, and uncovering the hidden machinations of political and economic systems can sometimes be disorienting. This creates the need for a clear structure and guideposts along the path to frequently reorient ourselves with the landscape and to stay focused on our primary mission: Build a technical, legal, and socioeconomic lifeboat that enables humanity to escape the problems described in this book.

Easy Navigation. Compasses and guideposts help explorers navigate treacherous environments. At times, it will feel like we are explorers on a nail-biting journey into uncharted psychological, philosophical, economic, and nonpartisan geopolitical territory. To maintain our bearing and to stay focused and mentally organized throughout this journey, each chapter ends with a section called "Key Points." This section summarizes the top-three principles in each chapter to maximize your retention of the most important concepts.

Unique Structure. The structure of this book is substantially different than most other books because it's intended to be an educational *and* practical handbook for busy people when they're discussing the principles herein with their friends, family and colleagues. To make it easy to navigate, I've added many page and paragraph headings throughout every chapter. This enables people to scan any page and quickly absorb the most important principles in seconds. This is especially useful for people who need to refresh their memory about a particular concept just before an important meeting or event.

Authentic Delivery. The tone of the human voice conveys approximately 30-40% of the total payload of meaning in verbal communications. This means reading a book without orienting the reader to the sound of the author's real voice can sometimes lead to misunderstood phrases, misinterpreted intentions, and less authentic human engagement with the material. For this reason, I will occasionally post videos discussing various topics in this book on the official Gini Foundation YouTube channel (YouTube.com/GiniFoundation). Watching and listening to those videos only takes a few minutes, but I think it will add a deeper human dimension as you read the rest of this book.

Scope & Focus

Global Perspective. I'm an American citizen and I frequently write nonpartisan articles about technical, economic and geopolitical issues related to American domestic and foreign policies. I've traveled to dozens of countries throughout this ever-shrinking world—including in Europe, Asia, the Middle East, Africa, North and South America—and I've studied other governments, economies, cultures, and languages as much as possible over the years. I've learned enough about this planet to confirm that the problems covered in this book are not limited to the United States. In fact, the same problems are unfolding in many countries around the world today.

Universally Applicable to Every Human. Although many of the examples and case studies in this book are focused on the U.S. political and economic systems, the solutions described in this book are applicable to every human on Earth today.

Nonpartisan, Non-Biased Perspective. Anybody who has read my books and articles knows that I write from a *nonpartisan and non-biased* perspective.[1] It doesn't matter if a politician's name is Trump, Bush, Clinton, Obama, Reagan . . . all their brains are hijacked by the broken incentives in Washington. Additionally, once you scratch beneath the surface of their shallow talking points, it's easy to see that *the outcomes* of the economic and foreign policies of nearly all national-level U.S. politicians since the early 1980s have been substantially identical. For all those reasons, and to avoid any appearance of partisan favoritism, I generally don't name individual politicians when discussing public policies because their names are merely incidental to the power of the political offices they control.

Focused on Broken Economic & Governance Systems, Not Individual People or Organizations. We are not attacking any particular person or organization in this book, but the purpose of public blockchain and cryptocurrency technologies is to resist various forms of government and corporate tyranny. So, Part 1 of this book covers several major systemic problems (especially regarding the U.S. Government) that are hurting humanity today. Whenever a particular person or organization is mentioned, it's entirely incidental to the larger systemic problem being discussed; i.e., their names are merely part of the back-story of the larger issue.

We Can't Please Everybody All the Time. Even if a divine angel glided down a gleaming shaft of light from the heavens and delivered this book directly, I'm sure several special interest groups would still find reasons to attack this book and me personally. Nevertheless, I suspect most conscious humans will appreciate it based on the feedback I've received so far. And if the billions of humans worldwide who are suffering from broken democracy and broken capitalism today are lucky, maybe their elected government officials will also read this book to help them solve many problems that they face today.

An Inspiring Journey. There is no satisfaction in the mere observation of tragedy. However, living a purposeful life can give us

---

1 The difference between "un-biased" and "non-biased" is important. For a summary of this distinction, please see: GiniFoundation.org/kb/why-trust-gini

opportunities to learn and share in ways that contribute meaningful solutions for the pain and suffering afflicting billions of humans on Earth today. For that reason, writing and sharing this book and the solutions herein is inspiring and gratifying for me and our Gini team. I hope our journey together in this book and beyond is inspiring and worthwhile for you, too.

My best,

Ferris Eanfar

Disclaimer. The usual disclaimer applies: The opinions expressed in this book are mine alone and may not necessarily reflect the opinions of any other person or organization.

Part I
The World Today

- Chapter 1 -
The System Is Broken

"Bad laws are the worst sort of tyranny."
— Edmund Burke

Introduction

Humanity Evolves at the Speed of Justice. Our technology is evolving at an exponential rate, but our laws and cultural norms evolve at *the speed of justice*. The speed of justice is dictated by the groups who have the most financial and political power in a society. The more concentrated the power, the slower the justice. Indeed, "justice" can only exist when economic and political systems produce outcomes that maximize health, wealth and political freedom for the largest number of humans. By that measure, the speed of justice in the United States and many countries on Earth today is excruciatingly slow. In fact, it's not only slow; it's accelerating *in reverse*.

To Appreciate Any Path to Justice, We Must First Recognize Injustice. This is a book about justice and the technologies, systems, and public policies that are necessary to achieve justice. To appreciate why justice is important, we should understand what is *unjust* about the status quo. The first draft of this book was over 500 pages because it included a detailed summary of the egregious economic and political injustice in our world today and hundreds of supporting scholarly third-party references. However, the size of the previous version made it unnecessarily tedious for readers to get through Part 1 to see the positive solutions in Part 2. So, in this version, I've condensed Part 1 to just a few chapters and I've placed much more emphasis on the solutions in Part 2.

10 Shocking Facts About Economic Power on Earth Today. To begin our brief exploration of the injustice on Earth today, consider the following facts.

- In the year 2000, the 100 largest economic entities on Earth were 51 corporations and 49 countries. In 2017, 69 are corporations and 31 are countries.[2] At the current rate, approximately 100 corporations will substantially control every material aspect of human life on Earth by 2045.
- The 10 largest corporations have more combined economic power than 92% (180) of all the countries on Earth *combined*.[3]
- The revenue of the 200 largest corporations is approximately 25% of Earth's GDP. In other words, less than 0.0002% of the corporations on Earth control 25% of humanity's income. The largest 2,000 (0.002%) corporations control 50% of humanity's income.[4]
- While consuming 50% of Earth's income, these 2,000 corporations employ less than 1% of Earth's labor force.
- The 50 largest financial corporations on Earth control over $70 trillion in assets, which was 91% of Earth's GDP in 2014.[5] (This concentration is certainly higher today.)
- Over 40% of all multinational companies on Earth are controlled by only 147 transnational cannibals (0.02% of all multinational companies); 75% of those companies are gigantic financial conglomerates, which have become *too big to fail;* 737 transnational cannibals (0.1% of all multinational companies) *control over 80% of all multinational corporations worldwide.*[6]

---

2 All data is derived from the CIA World Factbook and the Fortune Global 2000 List for the years indicated.

3 In this case, "financial power" is measured by annual corporate revenue for corporations and tax revenue for countries.

4 Derived from the 2014 Forbes Global 2,000 list and based on global GDP of $77 trillion. All the preliminary 2016 data indicates the concentration is even higher today.

5 Ibid.

6 See my previous book, *Broken Capitalism: This Is How We Fix It*, for a detailed explanation of what "transnational cannibal" means. The stats cited here are based on a rigorous analysis of 600,508 transnational corporations. However, these stats don't take into account the control that transnational cannibals have over global supply chains, political systems, industrial regulations, and market pricing. So, the actual control that these entities have is much greater than the figures cited here. For a detailed statistical analysis, see: Vitali, S., Glattfelder, J. B., & Battiston, S. (2011). The Network of Global Corporate Control. PLOS ONE, 6(10), e25995.

Continues on next page.

- The richest 1% of humans now have more wealth than the rest (99%) of the planet combined. And within that 1% group, the top-1% of 1% (the top-0.01%) own approximately 90% of all the 1%'s wealth. That means the majority of the 1% has more in common with the 99% than many people realize.[7]

- The eight richest humans own more wealth than the bottom 50% of Earth's entire human population.[8] The amount of wealth these eight humans *control* is even larger.

- More than 3 billion humans live on less than $2.50 per day. Approximately 5.5 billion humans (80% of the global population) live on less than $10 per day.[9]

- Approximately 50% of all children on Earth are living in poverty.[10]

The Essence of a Broken System. With such enormous economic and political power concentrated in the hands of less than 0.01% of Earth's corporations and less than 0.0000001% of Earth's human population, is it any surprise that economic and political systems worldwide are imploding, humanitarian crises are exploding, widespread protests and violence are raging, and 80% of the global population lives in poverty? Is this the "free market" that we learn about in college Economics textbooks and on mainstream financial news programs? All the problems discussed throughout this book are only possible when gigantic corporations and politicians collude against the best interests of

---

Doi.org/10.1371/journal.pone.0025995

7 Hardoon, Deborah, Ricardo Fuentes-Nieva, and Sophia Ayele. "An Economy For the 1%: How Privilege and Power in the Economy Drive Extreme Inequality and How This Can Be Stopped." Policy & Practice, January 18, 2016. Policy-practice.oxfam.org.uk/publications/an-economy-for-the-1-how-privilege-and-power-in-the-economy-drive-extreme-inequ-592643.

"Oxfam Says Wealth of Richest 1% Equal to Other 99%." BBC News, January 18, 2016, sec. Business. BBC.com/news/business-35339475.

"World's Eight Richest People Have Same Wealth as Poorest 50%." The Guardian, January 15, 2017, sec. Business. Theguardian.com/global-development/2017/jan/16/worlds-eight-richest-people-have-same-wealth-as-poorest-50.

8 Ibid.

9 If you want to learn more about global poverty, see the Global Issues website, which has aggregated poverty statistics from many primary sources: Globalissues.org/article/26/poverty-facts-and-stats

their citizens. This is the essence of broken capitalism and broken democracy on Earth today.

No Conspiracy Theories Needed. Many of the facts in this book are shocking, but this is not a book about conspiracy theories. We don't need conspiracy theories when the truth is already shocking enough; and the truth is sufficient to explain everything that is wrong with democracy and capitalism (the "System") on Earth today. In fact, I've written several books, over 100 articles, and the Global Governance Scorecard, documenting all the facts about how capitalism and democracy are dying in the United States and in several other countries today. This is all real and well-documented. Although I've stripped out most of the data and details from this book to make it easier to read, people are welcome to read the many articles and resources at GiniFoundation.org and Eanfar.org if they ever want more proof.

Broad Scope and Depth. I've spent many years researching, working, living, and writing about business, technology, and International Political Economy. So, I know it can take years to develop a broad and deep enough understanding to appreciate why the solutions in this book are more effective than other ideas discussed in policymaking circles today. To accelerate the learning process for my readers, this book succinctly summarizes a wide range of topics, including the most interesting and significant historical, cultural, technological, philosophical, and geopolitical facts and events that are necessary to understand why and how capitalism and democracy are broken in many countries today.

Concise Structure & Delivery. I don't have years to bring my readers up to speed on these topics. So, the structure of this book is designed to communicate the most important concepts in very potent, concise, informal language, without unnecessary fluff or fillers. My style is unorthodox, but I think this is the best way to achieve maximum readability, comprehension, and retention, with an occasional dash of humor and irony to keep things interesting.[11]

OK, without further ado, let's dive in. . . .

---

10 Ibid.

11 The "Navigation Tools / Structure" section in the Preface provides additional details about the
Continues on next page.

The Government vs. The Citizens

The "U.S. Government" is Not the "American People". When I say, "U.S. Government" (USG), I'm usually referring specifically to politicians and senior officials in Congress and the White House who create U.S. laws and public policies. It's important to differentiate the dysfunctional behavior of politicians and senior officials from all the civil service government employees and military personnel in the U.S. Armed Forces who have no control over the decisions of the USG. In fact, many USG employees are simply not aware of all the nasty things their political bosses have been doing in their name for generations.

The USG is Not Truly Representative of the American People. In addition to all the nonpartisan articles at Eanfar.org and GiniFoundation.org that confirm this, many nonpartisan studies have confirmed that the USG is controlled by a tiny handful of special interest groups whose interests and decisions are not aligned with the interests and decisions of the broader American public.[12] The concept of *democracy* can be inspiring in theory, but in practice, it simply does not exist in the United States. What *actually* exists in the U.S. today is a duopolistic political oligarchy with plutocratic tendencies that spawns perpetual corporate welfare programs, which politicians *sell* to the highest bidders in exchange for electoral campaign financing.

Governments Are Not Monolithic Monsters. The human brain creates models of the world and projects those models onto our minds in the form of human perception. Models are shortcuts that simplify reality, *but they are not reality.* This is problematic because virtually all political and economic ideologies are distorted by this biological phenomenon. For example, many people who have never worked inside a government institution falsely assume a "government" is a monolithic creature that thinks and acts with a unified purpose, which they assume is either malevolent or benevolent. This is a gross simplification that leads to all kinds of false assumptions, distorted perceptions, and unjust

---

unique structure of this book.

12 If you're not convinced, see the Gini Book List for many examples: GiniFoundation.org/kb/book-list

generalizations about the nature, quality, character, and intentions of the *individual humans* working in all the diverse departments within a government.

Distinguish Between Political Predators & Good-Faith Humans. A "government" is a diverse ecosystem of humans and human-driven processes, which are often dominated by a tiny number of *individual* political predators at the top of a political hierarchy. These predators (*not the "government"*) are to blame when governments hurt their citizens. There is nothing to fear from "government" because "government" is just an abstract mental shortcut that condenses thousands of diverse humans and human-driven processes into a single, simple label: "government."

The Real Threat Facing Humanity Today. The threat that liberty-loving humans *should* fear is *individual political predators* who hijack the organs and culture of government to serve their own self-interest and/or the interests of a tiny group of gigantic corporations and their largest shareholders. This distinction is important because it refines our perception of reality: The threat is not an all-powerful *monolithic government monster*, the threat is a small number of political predators in every generation who use their political power to support and symbiotically exploit a perpetually expanding, *unelected* global corporatocracy. Nearly every conceivable human-made problem on Earth today can be logically traced to this fundamental reality.[13]

It Can be Difficult for Non-Americans to See the Reality of the U.S. Political System. Until Donald Trump's 2016 presidential election, it was relatively difficult for humans living in other countries to understand how and why the U.S. political system is so broken. They often see the spectacle of U.S. presidential elections every four years in the media and then they assume that democracy in America must be working. In the past, I've often heard non-Americans say, "compared to the dictators in my country, American democracy looks like heaven." I don't hear that very much anymore. Why? Whether you like Trump or not, the circus that has engulfed his presidency has revealed the toxic

---

13 If you're not convinced, see the Gini Book List for many examples: GiniFoundation.org/kb/book-list

underbelly of the U.S. political system, which has been festering beneath the surface and degrading the integrity of the U.S. political system for generations.

Distinguish Between Political Theater and Political Reality. Humans around the world are finally beginning to see the reality: The U.S. political system is an elaborate theatrical show with many clever illusions. Prior to the Trump Era, this reality was difficult to perceive unless you lived inside the U.S. for several years and rigorously studied how the U.S. political system works beneath the surface. However, it is obvious to any American who has ever managed a substantial company, or conducted business in foreign markets, or been trapped by the USG's global diaspora tax, or tried to run for any national political office, among many other aspects of American life that are corrupted by the broken U.S. political system. This is why the USG ranks near the bottom of all OECD countries in the Global Governance Scorecard.[14]

We Are All Paying the Price for Trusting Political & Banking *Experts.* WWII, the Korean War, the Vietnam War, the Iranian Revolution, the 9/11 terrorist attacks, the 2003 Iraq War and ongoing humanitarian crises, the 2008 financial crisis and ongoing banking system implosion, the ongoing refugee crisis in Europe caused by the Iraq War . . . none of these events *needed* to happen. Those and many other tragedies have occurred *primarily* (not exclusively) because U.S. politicians and bankers (self-described public policy "experts") habitually ignore the obvious long-term consequences of their short-sighted and self-serving actions. Now, all of humanity is paying the tragic price for trusting the *experts.*

Injustice: Mass Murder & Endless Wars

At any moment, politicians in your country can pass lethal, liberty-killing, and financially devastating laws without your consent. The following list of laws and so-called *lawful actions* illustrates the harm that politicians cause when they give themselves the *lawful* power to inflict death, poverty and destruction upon millions or billions of humans.

---

14 See the Global Governance Scorecard report at Eanfar.org.

Genocide, Kidnapping, & Theft. The U.S. Indian Removal Act of 1830 gave the USG *lawful* power to slaughter, physically kidnap, steal their land, and expel millions of American Indians to tiny *Indian reservations*. During every slaughter, kidnapping, theft, and forced expulsion, the USG claimed it was a "wise and humane policy" *for the Indians' own benefit.*[15]

Concentration Camps. U.S. Public Law 503 gave the USG *lawful* power to send over 127,000 innocent American citizens to concentration camps throughout the 1940s.[16] Why? Because the Roosevelt Administration assumed Americans of Japanese ancestry would defect to the Japanese side during WWII. FDR's administration ignored the fact that nearly 70% of those innocent Americans were born and raised in the U.S. and had no political or cultural loyalty to Imperial Japan. Senator Robert Taft said, "I think this is probably the sloppiest criminal law I have ever read or seen anywhere. . . . I have no doubt that in peacetime no man could ever be convicted under it, because the court would find that it was so indefinite and so uncertain that it could not be enforced under the Constitution."[17] The USG did not acknowledge this atrocity until nearly 50 years later in 1988 when it offered a trivial $20,000 to each of the survivors. Of course, most of them were dead by then.[18]

Millions of Unnecessary Human Deaths. In 1964, U.S. President Johnson lied about USG intentions in Asia, lied about being attacked by the Vietnamese, and used the false attack story to persuade Congress to pass the Gulf of Tonkin Resolution. That gave Johnson *lawful* power to execute the largest air bombing campaign in world history; murder millions of Vietnamese, Cambodians, and Laotians; and unnecessarily sacrifice 58,220 American lives during the 20-year-long Vietnam War.[19]

---

15 Brands, H.W. (2006). Andrew Jackson: His Life and Times. Anchor. p. 488.

16 Yes, these really were "concentration camps," not "internment camps," as political operatives often like to call them. See "Terminology: Japanese American Incarceration and Japanese Internment." Densho.org/terminology.

17 Public Law 503 | Densho Encyclopedia. (n.d.). Encyclopedia.densho.org/Public_Law_503

18 Government Publishing Office. (1988). 100th Congress. Gpo.gov/fdsys/pkg/STATUTE-102/pdf/STATUTE-102-Pg903.pdf

19 Sheehan, N. (1989). A Bright Shining Lie: John Paul Vann and America in Vietnam (1st Vintage Books ed edition). New York: Vintage.

Endless Fear-Mongering. As they did in dozens of other countries, the USG falsely labeled Vietnamese civil rights activists as "communists" to create an excuse to bomb them. In reality, these "communists" had no interest in the ideology of communism; they were simply forced to choose between four bad options: (1) Continue to allow France and the USG to oppress them; (2) align themselves with the Soviets to gain their financial and military support; (3) align themselves with the Chinese to gain their financial and military support; or (4) fight like hell. They chose the last three options. But the Vietnamese revolutionary leader Ho Chi Minh was not a communist. In fact, during the early days of French domination in Vietnam he expressed admiration for the U.S., but Ho Chi Minh did not want his country to be culturally and economically raped and pillaged by *any* foreign nation. (Is that too much to ask?) So, he was forced to lead a fierce resistance movement (the Viet Cong) against the USG.

Endless *Democracy*-Bombing Campaigns. Air Force Chief of Staff Curtis LeMay perfectly summarized the USG's *solution* for civil rights activists when he was designing the USG's *lawful* bombing campaign: "We're going to bomb them back into the Stone Age."[20] Of course, preemptively bombing countries until they accept American-controlled *democracy* never works. As a result, Vietnam was the longest war in U.S. history, the USG dropped more bombs and chemical weapons than were used *by all nations combined* throughout World War II, costing American taxpayers over $1 trillion (in 2015 USD) for the Vietnam War, resulting in 2–4 million deaths. . . .[21] The USG's *lawful democracy-bombing* campaign continues today in the form of the so-called *War on Terror*, already resulting in more than triple the financial cost and approaching the same human cost of the Vietnam War.[22]

The Bogus "Domino Theory". Contrary to USG propaganda, Soviet and Chinese communist leaders were not seeking to colonize Vietnam during the 20th Century. In fact, they limited their involvement to providing supplies and weapons to the Viet Cong, which was

---

20 LeMay, C. E., & Kantor, M. (1965). Mission with LeMay; my story, p. 565. Garden City, NY: Doubleday.
21 See Sheehan (1989).
22 As of 2018, the never-ending war in Afghanistan is also the longest war in U.S. history.

primarily intended to deter further USG domination in their backyard. And after the war, of course, Indonesia, Singapore, Thailand, Malaysia and the Philippines did not *fall like dominoes* to the *communist menace* as the USG hysterically predicted. Why not? Because every country has its own unique political and cultural DNA, which cannot be easily mutated into pure communism *or* pure democracy. This is why, since the end of WWII, the USG's compulsive democracy-bombing adventures have failed in every case and why the fall of Saigon in 1975 did not trigger a communist domino effect throughout the world.[23]

What Have We Learned from This Section? These brief examples of *lawful* injustice, mass murder, and endless wars reveal that U.S. politicians will create dubious laws and frequently abuse their power for years or decades until millions of humans have died and USD trillions have been wasted. Thus, humans in the U.S. and around the world have many rational and legitimate reasons to distrust U.S. politicians and find ways to protect themselves from their toxic tyranny.

Injustice: Destruction of Liberty

Mass Surveillance. The U.S. Patriot Act of 2003 gave the NSA *lawful* power to conduct mass surveillance on every human on Earth, among other human rights violations. Most of the Patriot Act provisions are still in force today, despite the FBI admitting that the Patriot Act has not enabled them to prevent a single act of terrorism.[24] The U.K. Parliament has passed similar draconian *Terrorism Acts*, each of which tightens the noose around human rights more and more without addressing the fundamental causes of terrorism at all.[25,26]

---

23 The USG's bombing of Nazi Germany and Imperial Japan during WWII were legitimate defensive wars, not preemptive democracy-bombing adventures. Those countries became sustainable democracies after WWII because their populations recognized the legitimacy of the USG's defensive actions and the legitimacy of the USG's post-war reconstruction assistance. However, "legitimacy" has been absent in virtually all of the USG's military actions since WWII.
24 FBI admits Patriot Act snooping powers didn't crack any major terrorism cases. Washingtontimes.com/news/2015/may/21/fbi-admits-patriot-act-snooping-powers-didnt-crack
25 Horgan, J., & Horgan, J. (2013). U.S. Never Really Ended Creepy "Total Information Awareness" Program*. Blogs.scientificamerican.com/cross-check/u-s-never-really-ended-creepy-total-information-awareness-program
26 Terrorism Acts. (2018, March 28): Wikipedia.org/w/index.php?title=Terrorism_Acts

Tyranny Extended into the Afterlife. As incredible as this may seem, the Chinese "Reincarnation Law" passed in 2007 literally requires Tibetan Monks to get permission from the Chinese Government before they can die and be reincarnated, which extends *lawful* government tyranny into the afterlife.[27]

Taxing Your Air & Sunshine. The British "Window Tax" was literally a *lawful* tax on air and sunlight, which was passed in 1696 and taxed British *subjects of the Crown* (i.e., British citizens) for over 150 years.[28] Imagine being taxed for over 150 years simply for having windows that provide air and sunshine.

Imprisonment & Deportation for Anti-War Speech. The U.S. Alien and Sedition Acts were a series of laws passed in 1798, which criminalized anti-war speech and gave the president *lawful* power to deport non-American *legal* residents who were critical of the federal government. Most of these Acts were repealed by Thomas Jefferson, but the Alien Enemies Act has lingered in various forms and has been the legislative basis for *lawfully* incarcerating innocent Americans for anti-war speech during times of war *and* during undeclared wars for over 200 years.[29]

Destruction of Free Speech & Independent Journalism. The 1917/1918 U.S. Sedition and Espionage Acts made anti-war speech a crime (*again*). Although these Acts have been amended, the Espionage Act is still used today to *lawfully* silence journalists like Daniel Ellsberg and Glenn Greenwald, among others, who have exposed systemic government corruption.[30,31]

Government-Mandated Racial Prejudice & Segregation. Even after slavery was abolished in 1864, the Jim Crow Laws and U.S. Black

---

27 Yes, this law really does exist. See:
Wikipedia.org/wiki/State_Religious_Affairs_Bureau_Order_No._5
28 Glantz, A. E. (2008). A Tax on Light and Air: Impact of the Window Duty on Tax Administration and Architecture, 1696-1851, 24.
29 The Alien and Sedition Acts - Constitutional Rights Foundation. (n.d.). Crf-usa.org/america-responds-to-terrorism/the-alien-and-sedition-acts.html
30 Greenwald, G. (2013, June 22). On the Espionage Act charges against Edward Snowden | Glenn Greenwald. Theguardian.com/commentisfree/2013/jun/22/snowden-espionage-charges
31 What would it take to nail Glenn Greenwald under the Espionage Act? - The Washington Post. (2013, June 25). Washingtonpost.com/blogs/erik-wemple/wp/2013/06/25/what-would-it-take-to-nail-glenn-greenwald-under-the-espionage-act/?noredirect=on&utm_term=.204bf90982cf

Codes *lawfully* restricted the rights and liberties of African Americans for over a century until they were finally repealed in 1965.[32]

Legislating Private Morality. The 18th Amendment of the U.S. Constitution banned "the manufacture, sale, and transportation of alcoholic beverages. . . ." As the so-called *War on Drugs* and the Prohibition Era prove, trying to legislate what humans put into their own bodies based on highly subjective moral ideologies is a losing and costly battle for any democratic society. The Temperance Movement, including the Anti-Saloon League and many religious organizations, had wanted to ban alcohol for over 100 years prior to the 18th Amendment's passage in 1919. Of course, they had good intentions, but good intentions are never enough. The politicians who meekly capitulated for political reasons to the vocal minority of anti-alcohol advocates *lawfully* ignited the most violent and destructive crime wave in American history.[33] It also gave enormous power and wealth to Al Capone and other crime bosses, gave J. Edgar Hoover emperor-like power over the U.S. political system for decades, and transformed the FBI into a tyrannical domestic spy agency that was just as corrupt (though not as blood-thirsty) as Stalin's NKGB.[34]

Exploiting the Masses to Make the Rich Even Richer. The British Corn Laws were passed in 1815, which *lawfully* made the price of corn artificially high so that landowners (the most powerful special interest group) could enjoy artificially high profits for over 30 years.[35] Any landowner that grew corn could sell it for about 100 times higher than the price of corn today.[36] Naturally, this angered British *subjects of the Crown* and they revolted. The British Parliament deployed the military to suppress the revolt, which resulted in the Peterloo Massacre. The British Parliament then passed the draconian Six Acts laws to *lawfully* suppress free speech, silence newspapers, prevent reform activists from organizing meetings, and shutdown any form of public resistance to the

---

32 Southern Black Codes - Constitutional Rights Foundation. (n.d.). Crf-usa.org/brown-v-board-50th-anniversary/southern-black-codes.html
33 Burrough, B. (2005). Public Enemies: America's Greatest Crime Wave and the Birth of the FBI, 1933-34. New York: Penguin Books.
34 Ibid.
35 Corn Laws. (n.d.). Princeton.edu/~achaney/tmve/wiki100k/docs/Corn_Laws.html
36 About 10 times higher on a PPP basis.

massacre and the lack of voting rights for women. The last of these six laws was not repealed until 2008, *nearly 200 years later.*[37]

Blocking Technological Innovation. The U.S. and U.K. Red Flag Laws during the late 1800s required automobile owners to walk in front of their cars (defeating the purpose of an <u>auto</u>mobile) and wave red flags *allegedly* so they would not scare people, horses and cows. Then, if they happened to see a person, horse or cow, the automobile owners were required to (1) immediately stop the vehicle; (2) "immediately and as rapidly as possible . . . disassemble the automobile"; and (3) "conceal the various components out of sight, behind nearby bushes" until the people, horses and cows were gone.[38]

What Have We Learned from This Section? These brief examples of injustice and the destruction of our liberty reveal that politicians will create dubious laws and frequently abuse their power for years or decades until millions of humans have suffered and USD trillions have been wasted. Thus, we have many rational and legitimate reasons to distrust politicians and to find ways to protect ourselves from their tyranny.

Injustice: Destruction of Wealth

Mass Economic Destruction Spawns War. Despite the dire warnings of over 1,000 economists, the U.S. Smoot-Hawley Tariff Act of 1930 was passed. The Act raised tariffs on over 20,000 goods by up to 20%. This amplified the effects of the 1929 stock market crash and transformed a recession into a decade-long, worldwide Great Depression. The suffering was so bad that it enabled people like Adolf Hitler in Germany and the radical militarists in Imperial Japan to seize political power, which ignited World War II and caused over 70 million deaths.[39,40]

---

37 BBC. (2008). Peterloo law set to be repealed.
BBC.co.uk/manchester/content/articles/2008/03/19/190308_peterloo_law_feature.shtml
38 Olyslager, P. & Sir J. Brabham. Illustrated Motor Cars of the World. New York: Grosset & Dunlap, 1967.
The likely real reason was to protect special interest groups like horse-and-buggy producers from the proliferation of automobiles.
39 Eanfar, Ferris. What Caused the Great Depression? (Part 1). (2016, July 12).
Continues on next page.

Massive Financial Crisis. The U.S. Gramm-Leach-Bliley Act (GBA) was *purchased* by the banking lobby for approximately $300 million. GBA repealed the Glass-Steagal Act, which enabled commercial banks to *lawfully* gamble with depositor assets, thereby creating greater systemic economic instability. The 2008 Financial Crisis was the inevitable result.[41]

Worldwide Destruction of Wealth. The politically motivated U.S. Community Reinvestment Act distorted the incentives throughout the global financial system because of the distortions it injected into the U.S. mortgage lending industry and all the worldwide mortgage-backed securities associated with it. This pushed the global economy into the 2008 Financial Crisis, which resulted in at least $29 trillion in lost wealth.[42] This has *lawfully* accelerated the destruction of the middle class throughout the Western world. The corrupt policies that spawned the 2008 crisis are still plaguing Europe and many Americans to this day.

Empowering Drug Companies to Exploit Americans. The U.S. Bayh-Dole Act (1980) and Hatch-Waxman Act (1995) give pharmaceutical companies cheap access to valuable intellectual property (IP) created at public universities funded by taxpayer dollars.[43] Then they wrap their marketing around that IP and sell harmful and/or unproven drugs back to Americans at triple- and quadruple-digit profit margins.[44] Then Americans are taxed again to subsidize an egregiously broken healthcare system that provides rubber-stamped, non-negotiable, highly profitable contracts to the drug companies.[45] Thus, American taxpayers get *lawfully* taxed and gouged three times to subsidize a $400

Eanfar.org/caused-great-depression-part-1

40 Eanfar, Ferris. The Birth of the Modern Welfare State. (2017, August 31). Eanfar.org/birth-modern-welfare-state

41 See my previous book, Broken Capitalism: This Is How We Fix It, which contains hundreds of independent references and charts to illustrate the consequences of broken economic policies.

42 Ibid. and Carney, J. (2011, December 14). The Size of the Bank Bailout: $29 Trillion. Retrieved April 20, 2017. CNBC.com/id/45674390

43 Jones, G. H., Carrier, M. A., Silver, R. T., & Kantarjian, H. (2016). Strategies that delay or prevent the timely availability of affordable generic drugs in the United States. Blood, blood-2015-11-680058. Doi.org/10.1182/blood-2015-11-680058

44 Angell, M. (2005). The Truth About the Drug Companies: How They Deceive Us and What to Do About It (Reprint edition). New York: Random House Trade Paperbacks.

45 Eanfar, F. (2018).Global Governance Scorecard.

billion industry that spends far more money on deceptive marketing campaigns than on real R&D.[46]

Confiscating Our Wealth. U.S. Executive Orders 6102 (April 5th, 1933) and 6814 (August 9, 1934) gave the U.S. Government the power to *lawfully* confiscate all the gold and silver owned by American citizens.[47] Similar laws have existed in Australia and India.[48,49] Then there's the "bail-in" that occurred in Cyprus in 2013 when the Cyprus Government allowed the banks to *lawfully* and literally steal 50% of their citizens' money.[50] This bail-in tactic is now an official policy in several countries today.[51,52] For example, Canada's Economic Action Plan 2013 explicitly states:

The Government proposes to implement a "bail-in" regime for systemically important banks. This regime will be designed to ensure that, in the unlikely event that a systemically important bank depletes its capital, the bank can be recapitalized and returned to viability through the very rapid conversion of certain bank liabilities into regulatory capital. This will reduce risks for taxpayers. The Government will consult stakeholders on how best to implement a bail-in regime in Canada.

---

46 Swanson, A. (2015, February 11). Big pharmaceutical companies are spending far more on marketing than research. Washington Post.
Washingtonpost.com/news/wonk/wp/2015/02/11/big-pharmaceutical-companies-are-spending-far-more-on-marketing-than-research
47 FDR Confiscates Gold from Citizens: Executive Order 6102. (1933, April 5). U.S. Government Printing Office. Goldline.com/images/conf-order.pdf
48 Franklin D. Roosevelt Executive Order 6814—Requiring the Delivery of All Silver to the United States for Coinage. (1934). Presidency.ucsb.edu/ws/index.php?pid=14741
49 Suchecki, Bron (August 4, 2008). "A History of Gold Controls in Australia".
50 The Confiscation of Bank Savings to "Save the Banks": The Diabolical Bank "Bail-In" Proposal | Global Research - Centre for Research on Globalization. (n.d.). Globalresearch.ca/the-confiscation-of-bank-savings-to-save-the-banks/5329411
51 It Can Happen Here: The Bank Confiscation Scheme for US and UK Depositors | Global Research - Centre for Research on Globalization. (n.d.). Globalresearch.ca/it-can-happen-here-the-bank-confiscation-scheme-for-us-and-uk-depositors/5328954
52 Lewis, N. (2013). The Cyprus Bank "Bail-In" Is Another Crony Bankster Scam. Forbes.com/sites/nathanlewis/2013/05/03/the-cyprus-bank-bail-in-is-another-crony-bankster-scam

Politicians & Banks Can Now Steal Our Fiat Bank Deposits. I've seen a few people say things like, "Taking depositors' money is not what the Canada action plan means. . . ." So, let's eliminate any doubt about what the statement above actually means: In the banking industry, the only "liability" that can be "rapidly converted" into "regulatory capital" is *customer deposits*. That's the definition of a "bail-in"; banks have no other "liabilities" to confiscate that can be converted into "regulatory capital." For people who understand banking, those words can only mean one obvious thing: The banks in Canada and several other countries have arbitrarily given themselves *lawful* power to steal their citizens' money whenever their politicians think it's necessary.

"Too Big to Fail" Is the Ideology that Dominates the Banking System Today. It's difficult for many people to believe that banks could steal their money because they don't really grasp how the global banking system has been hijacked by a tiny group of politicians and gigantic corporations guided by a self-serving ideology: "too-big-to-fail." This ideology has been *lawfully* injected like a virus into many governments on Earth today. That ideology philosophically sacrifices individual human liberty and ecosystem sustainability to the interests of the wealthiest shareholders and executives of the banks, large corporations, and politicians who keep *lawfully* making the same destructive mistakes over and over again.

Official **Tyranny is Just the Tip of the Iceberg.** So far, we've covered a small sample of the corrupt and/or deeply flawed laws that have been *officially* implemented without majority citizen consent, which have destroyed the wealth, liberty, and lives of hundreds of millions of innocent humans worldwide. Yet, those atrocious laws do not include the countless instances of human rights violations, regulatory corruption, and corporate welfare that have occurred in the shadows, which often go undetected by the general public for years or decades. Usually, targeted and/or marginalized individuals and groups suffer in silence and our collective human rights and wealth are continuously degraded by the collusion of gigantic corporations and governments one backroom deal and secret decision at a time.[53,54,55,56]

---

53 Schneier, B. (2016). Data and Goliath: The Hidden Battles to Collect Your Data and Control
Continues on next page.

Corporations Control & Corrupt U.S. Foreign Policy. In addition to everything covered so far, my article, "Is U.S. Foreign Policy Controlled by Corporations?" provides a relatively detailed historical summary of the collusion between the USG and gigantic U.S. corporations, which have extracted USD trillions of wealth from humans in the U.S. and around the world.[57] Additionally, below are just a few examples of domestic corporate and government collusion, which have caused millions of humans to suffer worldwide and USD trillions of wealth to be sucked from American taxpayers.

Creating Environmental Pollution & Terrorism. We know from congressional testimony that generations of American politicians secretly *colluded with* auto manufacturers and oil companies to *lawfully* block Thomas Edison from selling electric engines during the early decades of the 20th Century.[58] In fact, the auto and oil industries hijacked the regulatory apparatus of the USG by paying off politicians and funding their political election campaigns. In return, the politicians regulated Thomas Edison's electric vehicle technologies to death and allowed the big auto and oil companies to violate anti-trust laws and engage in monopolistic practices that prevented Edison from competing in the marketplace. This has stifled innovation and delayed the broad adoption of alternative energy technologies for over 100 years. It has also created the toxic dependence on Middle Eastern oil, terrorism, and geopolitical quagmires that our planet suffers from today.

The Biggest Scandal in Medical History. The cholesterol-lowering drug Lipitor has generated over $140 billion for the

Your World (1 edition). New York London: W. W. Norton & Company.

54 Greenwald, G. (2015). No Place to Hide: Edward Snowden, the NSA, and the U.S. Surveillance State (Reprint edition). New York: Picador.

55 Proof of Google Censorship & a Path to Freedom. Eanfar.org/proof-google-censorship-path-freedom

56 Greenwald, G. (2014, September 5). The U.S. Government's Secret Plans to Spy for American Corporations. Theintercept.com/2014/09/05/us-governments-plans-use-economic-espionage-benefit-american-corporations

57 Eanfar, F. Is U.S. Foreign Policy Controlled by Corporations? (2018, May 21). Eanfar.org/is-u-s-foreign-policy-controlled-by-corporations

58 Black, E. (2015). Internal Combustion: How Corporations and Governments Addicted the World to Oil and Derailed the Alternatives by Edwin Black. St. Martin's Press.

pharmaceutical company Pfizer.[59] And that's just one drug. During the same reporting period, the top-20 best-selling drugs generated nearly $1 trillion in revenue for a small handful of huge corporations. With income equivalent to entire countries, these corporations have a lot to lose if their products were ever proven to be ineffectual. Despite their best efforts to suppress this information, there's a growing body of peer-reviewed medical literature on the bad science behind these drugs.[60] Not only are they based on questionable data, *independent* research studies (i.e., research not directly or indirectly funded by drug companies) have revealed how dangerous these drugs really are.[61,62,63,64]

Rampant Fraud & Abuse in the Medical Industry. Marcia Angell, M.D., who was the Editor-in-Chief of the New England Journal of Medicine for over 20 years, said the drug companies were "illegally overcharging Medicaid and Medicare, paying kickbacks to doctors, engaging in anti-competitive practices, colluding with generic companies to keep their generic drugs off the market, illegally promoting drugs for unapproved uses, engaging misleading direct-to-consumer advertising, and, of course, covering up evidence. . . . "[65] If this was a truly democratic country, it would be impossible for fraud on this scale to exist for so long, *lawfully* sucking our wealth and quality of life downward into a bottomless healthcare black hole.

---

59 King, S. (2013). The Best-Selling Drugs Since 1996 - Why AbbVie's Humira Is Set To Eclipse Pfizer's Lipitor. Forbes.com/sites/simonking/2013/07/15/the-best-selling-drugs-since-1996-why-abbvies-humira-is-set-to-eclipse-pfizers-lipitor/

60 T. Demasi, D. M., & statins. (2014, May 13). The Cholesterol Drug War: ABC Australia Bans Documentary Exposing Statin Drug Scandal. Healthimpactnews.com/2014/the-cholesterol-drug-war-abc-australia-bans-documentary-exposing-statin-drug-scandal

61 Demasi, M. (2018). Statin wars: have we been misled about the evidence? A narrative review. British Journal of Sports Medicine, bjsports-2017-098497.

62 O'Connor, A. (2017, December 21). How the Sugar Industry Shifted Blame to Fat. The New York Times. Nytimes.com/2016/09/13/well/eat/how-the-sugar-industry-shifted-blame-to-fat.html

63 Bowden, J., & Sinatra, S. (2012). The Great Cholesterol Myth: Why Lowering Your Cholesterol Won't Prevent Heart Disease-and the Statin-Free Plan That Will (1 edition). Beverly, MA: Fair Winds Press.

64 How statin drugs really lower cholesterol & kill you one cell at a time – Zoë Harcombe. (n.d.). Zoeharcombe.com/2013/10/how-statin-drugs-really-lower-cholesterol-and-kill-you-one-cell-at-a-time

65 Angell, M. (2005). The Truth About the Drug Companies: How They Deceive Us and What to Do About It (Reprint edition). New York: Random House Trade Paperbacks.

The $1.5 Trillion Airplane & the Insatiable Military-Industrial Complex. After a military weapon (e.g., the $1.5 trillion F-35 Fighter Jet that "broke the U.S. Air Force") is contracted to be built by a defense company, politicians have strong political incentives to *find* reasons to justify the cost of the weapon, or else they won't get re-elected; and the weapons companies, consultants, and their lobbyists have strong commercial incentives to justify the cost of their existence.[66] Both of these symbiotic incentives converge to substantially inflate the significance of various global threats. These toxic incentives also inflate the amount of American taxpayer dollars that Congress allocates to the U.S. defense budget, which inflates the national debt and steals wealth from American citizens. These toxic incentives corrupt and distort U.S. national security policies and encourage U.S. politicians to:

- *Find (and fabricate)* **terrorist threats lurking in every corner of the planet.** For example, the entire war in Iraq was based on a single "*Nigerian* yellow cake uranium transaction" that never actually occurred, which U.S. politicians at the highest level tried to cover up.[67]
- Maintain a nuclear arsenal that is nearly 1,000 times larger than necessary to defend the U.S. homeland and U.S. allies.
- **Implement mass surveillance programs designed to spy on American citizens** via phone systems, banking systems, Internet systems, snail-mail systems, satellite systems, drones, and a sprawling network of law enforcement and intelligence agencies.
- **Create constitutionally dubious *fusion centers* in every state in America,** which are secret offices in American cities designed to spy on American citizens.[68]
- **Brainwash Americans with expensive propaganda,** e.g., *outspending the Soviet Union was the reason the U.S. won the Cold War.* In

---

66 Francis, D. (2014, July 31). How DOD's $1.5 Trillion F-35 Broke the Air Force. CNBC.com/2014/07/31/how-dods-15-trillion-f-35-broke-the-air-force.html

67 Yellowcake forgery - SourceWatch. (n.d.). Sourcewatch.org/index.php/Yellowcake_forgery

68 Kayyali, D. (2014, April 7). Why Fusion Centers Matter: FAQ. Eff.org/deeplinks/2014/04/why-fusion-centers-matter-faq

reality, the Soviet Union was corroding from the inside for many other reasons and the USG could have spent 1/50th of the amount on nuclear weapons and the Strategic Defense Initiative (aka the *Star Wars Program*) and the USSR would have still collapsed during the early 1990s.

What Have We Learned from This Section? These brief examples of injustice and wealth destruction reveal that U.S. politicians and gigantic corporations will create dubious laws and frequently abuse their power for years or decades until USD trillions of wealth has been wasted and destroyed. Additionally, we have learned that the collusion between politicians and gigantic corporations results in the widespread *redistribution* of wealth out of the middle class and into the hands of a small number of gigantic corporations, which use their wealth to make their executives, shareholders, and pliable politicians very rich. Thus, we have many rational and legitimate reasons to distrust politicians and to find ways to protect ourselves from their toxic tyranny.

The Clown Bus & the Cliff

When I was younger, I was enamored by the grandeur and rich history of the awe-inspiring historical monuments in Washington, D.C. Then when I was a bit older, I served in the U.S. Air Force because of my deep respect for George Washington, Thomas Jefferson, John Adams, and the ancient Athenian principles of democracy that made the United States the most prosperous and inspiring nation in the history of humankind. It was easy to be proud of America before I learned how far the U.S. political system had fallen from the ideals envisioned at her birth.

Now, all I can feel is the kind of steely melancholy that accompanies the obligatory tolerance one has for an alcoholic father who refuses to sober up. Instead of reverence and admiration, every time I hear the words "Congress," "White House," or "government," I often feel like I'm a passenger in an absurdly distorted bus speeding toward a cliff, driven by intoxicated harlequins with grotesque faces. Their mouths and bodies twitch and convulse at random intervals, but their words are meaningless. There is an invisible barrier between the American people in the passenger seats and the theatrically elected harlequins crammed into the driver seat. That barrier is the collective paralysis of a nation in decline, speeding toward a cliff.

Key Points

- **Anti-Democratic War-Mongering Never Creates Sustainable Peace.** USG officials often ignore the fact that democracy-bombing adventures in foreign countries create concussion blasts that impact the American homeland. This blowback destroys the integrity of domestic and foreign institutions and fuels distrust and hatred for the USG (and authority in general) in domestic and foreign populations.
- **Anti-Democratic Wealth Destruction Never Creates Widespread Prosperity.** The USG's compulsive destruction of our wealth by colluding with self-serving corporations has created the most indebted nation in human history. Hundreds of millions of Americans are now trapped in a life-long, multi-generational debtor's prison.
- **Anti-Democratic Political Systems Never Create Widespread Freedom.** Large populations don't tolerate economic and political oppression forever. The destruction of our wealth and liberty is not the result of democracy; it is the result of an ongoing and expanding tyranny. A deep feeling of injustice is boiling beneath the surface in many countries today, which so-called *elites* usually don't see until events like Brexit and Donald Trump's presidential election give them a dose of reality. Regardless of Trump's performance, his election is a reflection of the deep corrosion, anger and corruption that is destroying the American political system today.

- Chapter 2 -
The Surveillance State

"They who can give up essential liberty to obtain
a little temporary safety deserve neither liberty nor safety."
— Benjamin Franklin

"I See You." Visualize a man staring at you through your bedroom window. His cold, menacing eyes scan every inch of your body. He reaches into his pocket. Then your phone rings. "I see you," a sinister voice growls. Now, is your instinct to assume the man has good or bad intentions? Are you happy he is staring at you? Why not, are you a criminal? Most rational people would be uncomfortable in that situation. Why? Because all forms of spying and surveillance inhibit free speech, creative thinking and expression, and every human impulse and action. When people, governments, and corporations are staring at you, they control important aspects of your existence. The more power they have, the more control they have over you.

"If You Have Nothing to Hide, You Have Nothing to Fear." — Joseph Goebbels, Nazi Propaganda Minister. If it wasn't so disgusting, it would be amusing to see the veritable parade of politicians and corporate executives since 2001 parroting Goebbels, not realizing they're quoting one of the most notorious mass murders in human history as they contrive to justify their ubiquitous surveillance systems. Whenever I hear their Goebbels-inspired mantras, I usually say, "OK, please give me the passwords to your bank accounts, email accounts, social security number, and let me install an Internet camera in your bedroom and bathroom."[69] Of course, they will never give you access to

---

69 Investigative journalist Glenn Greenwald takes a similar approach, which he mentioned in his excellent book, No Place to Hide: Edward Snowden, the NSA, and the U.S. Surveillance State. That's one of the best books on this topic.

that private data because they know that would force them to behave in an unnatural, uncomfortable, and unfulfilling way from that day forward.

Former CIA employees agree that those who claim they have "nothing to hide" are dangerously naive. According to Suzanne Spaulding, former CIA Senior Attorney:

So many people in America think this does not affect them. They've been convinced that these [surveillance] programs are only targeted at suspected terrorists. [They think] "I'm not engaged in any terrorist activities; therefore, this does not concern me. There's no way in which I'm going to be caught up in this activity. . . ." It is inevitable that totally innocent Americans are going to be affected by these programs.[70]

Mass Surveillance Is Censorship. If you're living a quiet life that doesn't disrupt the flow of money and power to gigantic corporations and politicians, then you might be able to live and die in obscurity without any of them caring about your existence. But, if you're Mahatma Gandhi, Nelson Mandela, Martin Luther King, John F. Kennedy, Elizabeth Cady Stanton, Thomas Jefferson, James Madison, Harriet Tubman, Mohammad Mosaddegh, Patrice Lumumba, Jacobo Árbenz, Edward Snowden, Glenn Greenwald, and thousands of others in the past and present who courageously resist destructive and/or corrupt organizations to defend your human rights, then spying and surveillance is a very real threat. In fact, spying and surveillance is the single-biggest threat, short of physical death, because spying gives misguided politicians and organizations the power to disrupt and censor your words and thoughts *before you have a chance to communicate and act upon them.*

Privacy Is Protected in the Constitution. Some people have claimed that there is nothing in the U.S. Constitution about privacy. That's a nonsensical and highly disingenuous claim. The Constitution, and the 4th Amendment in particular, provide a sphere of privacy around each citizen, which cannot be penetrated without probable cause and a legally authorized warrant to search *a specific* person within *a specific*

---

70 Interview in a June 10, 2013 Frontline Investigative report

time frame at a *specific location*. There is nothing in the Constitution that gives the USG the authority to view, collect, or access every communication and transaction that occurs between private parties outside those very narrow conditions. In fact, the 4th Amendment clearly forbids mass surveillance of any kind.

Privacy Is a Fundamental Human Right. Regardless of whether the word "privacy" is codified in any national constitution, privacy is a fundamental human right.[71] In fact, the act of free speech—our most important human right upon which all other rights depend—is merely the last link in a long chain of events that enables a human to learn and communicate anything. If we are not free to privately explore controversial ideas, free to experiment privately to test our ideas, free to engage in commerce privately to obtain the items required to test our ideas, and free to live privately without a suffocating blanket of surveillance, then we will never be free to effectively use *any* of our rights to confront the powerful people, governments, and corporations that are destroying our economic and political systems today. If we give them the power to spy on us, we give them the power to destroy all the pillars of democracy and legitimate national governance, upon which a democratic civilization depends.

The National Security State

When Ben Franklin said, "They who can give up essential liberty to obtain a little temporary safety deserve neither liberty nor safety," he was speaking as one of the most accomplished and beloved statesmen in human history. Among many other accomplishments, Franklin served as Governor of Pennsylvania, as U.S. Ambassador to numerous countries, and he contributed to all the most pivotal debates during the development of the U.S. Constitution. With all his real-world experience protecting American interests against the tyranny of the British Empire, which was a far greater threat to American interests during the American Revolution than any terrorist group is today, why didn't Ben Franklin recommend the Patriot Act in 1776? Instead of his famous quote above,

---

71 Strasser, M. R. (2008, July 16). Fourth Amendment. Law.cornell.edu/wex/fourth_amendment

why didn't he say, "Without security, there is no liberty; thus, we must spy on all Americans"?

Liberty Is the Source of National Security. Franklin, Jefferson and all the U.S. Founders understood a fundamental truth about national security that many politicians ignore today: *Liberty* for our citizens and *respect for the liberty of other humans and nations* is the source of enduring security, *not* an ubiquitous police state. As we learned in chapter one, we have had short-sighted (and often corrupt) politicians in control of the USG for generations who have trampled upon the human rights of millions of humans and dozens of countries since the end of World War I.[72] When a government continuously and gratuitously invites the enmity of other nations, attacks on its people and homeland are inevitable. That's not a national security problem; that's a leadership and institutional integrity problem. *Those* problems threaten American liberty, democracy, the long-term interests of Americans, and the future of humanity far more than any terrorist group today.

Self-Inflicted Terror Does Not Legally Justify the Mass Destruction of Liberty. The USG's short-sighted foreign policy has amplified the dangers associated with terrorism since the 1970s, but that danger is self-inflicted. The USG's self-inflicted terror creates no moral or legal obligation for American citizens to give up their human rights merely because an entrenched political cartel in Washington habitually engages in hostile actions against foreign countries and incites worldwide violence against Americans. Thus, any suggestion that Americans must give up their human right to privacy and be condemned to a virtual panopticon cage to support an illegitimate mass surveillance state and illegitimate wars is manifestly illogical, unconstitutional, anti-democratic, and tyrannical.

---

72 Grossman, Z. (n.d.). From Wounded Knee to Syria. Sites.evergreen.edu/zoltan/interventions

Total Information Awareness. From the perspective of people who think the *War on Terror* and a mass surveillance state is a rational response to the 9/11 attacks, the concept of *Total Information Awareness* (TIA) is the holy grail of national security. TIA is a useful lens through which to understand their worldview and their goals because TIA is both a virulent ideology and an official surveillance program of the U.S. Information Awareness Office (IAO).[73] The TIA program was created in early 2003, but it was later renamed to "*Terrorism* Information Awareness" to de-emphasize the reality of its Orwellian nature. Nevertheless, the essence of its purpose ("Total Information Awareness" of all American citizens) never changed. In fact, the New York Times reported in 2012 that, despite the name change, TIA was "quietly thriving."[74]

---

73 Weinberger, Sharon (24 January 2008). "Defence research: Still in the lead?". Nature. Nature Publishing Group. pp. 390–393. doi:10.1038/451390a.

74 Harris, Shane (August 22, 2012). Giving In to the Surveillance State. New York Times. p. A25.

Perpetually Lying to Citizens. After senior U.S. and British politicians lied and deceived their citizens about the reasons for going to war in Iraq, later in 2003, senior U.S. politicians lied again and said the TIA and IAO were de-funded after unflattering press created political pressure to do PR damage control.[75,76,77,78] However, we know from the Snowden files released 10 years later that U.S. politicians did not reduce the scope of their TIA activities; in fact, they dramatically expanded them. We also know that numerous other USG officials lied under oath, in congressional testimony, and in all their public statements to conceal the truth of their actions and intentions.[79] When politicians systematically and perpetually lie, deceive, and distort reality while they allow gigantic security corporations to *lawfully* extract USD trillions in taxpayer wealth, we have no rational justification to trust that politicians are acting in the general public's best interest.

Do We Really Want Politicians to Be Omnipotent? The ideology of TIA is rooted in an intoxicating and deceptively simple idea: If politicians collect *enough* data, they could *theoretically* prevent any crime, anywhere, just like in the dystopian movie The Minority Report. To the extent that politicians, NSA, FBI, and CIA spies (and their counterparts in other countries) become omnipotent and omniscient gods, this is technically true. However, given the systematic deceptions, human rights violations, and self-righteous tyranny perpetrated by many politicians throughout human history, the prospect of politicians becoming omnipotent and omniscient is more frightening than anything George Orwell ever imagined. So, many humans are justifiably outraged by the

---

75 Chilcot report: key points from the Iraq inquiry. The Guardian. (2016, July 6).
Theguardian.com/uk-news/2016/jul/06/iraq-inquiry-key-points-from-the-chilcot-report

76 Schwarz, J. (2016, February 18). Trump Is Right, Bush Lied: A Little-Known Part of the Bogus Case for War. Theintercept.com/2016/02/18/trump-is-right-bush-lied-a-little-known-part-of-the-bogus-case-for-war

77 Allen, E. (2016, June 28). Chilcot Inquiry: What is it and what did the Iraq War report say? The Telegraph. Telegraph.co.uk/news/2016/06/28/chilcot-inquiry-when-is-the-report-being-published-and-why-has-i

78 Board, T. E. (2018, January 20). Opinion | Iraq War Lies, 13 Years Later. The New York Times. Nytimes.com/2016/07/08/opinion/iraq-war-lies-13-years-later.html

79 Lawson, S. (2013). Did Intelligence Officials Lie to Congress about NSA Domestic Spying? Forbes.com/sites/seanlawson/2013/06/06/did-intelligence-officials-lie-to-congress-about-nsa-domestic-spying

ever-expanding global surveillance state, regardless of the intentions of politicians and political officials who claim to be using their power to protect us from terrorism.

A State of National *Insecurity*

Mass Surveillance Destroys the Health & Integrity of Societies. When governments spy on their citizens, they create a prison of the mind for those who know they're under surveillance. Several studies have confirmed that, when innocent humans are under surveillance, fear and insecurity take control of their minds, which causes their mental and physical health to deteriorate.[80,81,82,83] They suffer from higher anxiety, which increases their susceptibility to cancer.[84] They no longer trust authority because they don't feel authorities trust them, which leads to a state of widespread anomie that causes the social bonds of a society to breakdown.[85,86,87,88] They automatically conform their behavior to whatever they think the government wants, not what is best for themselves and their society.[89] This becomes problematic when the self-

80 Stanton, J. M., & Barnes-Farrell, J. L. (1996). Effects of electronic performance monitoring on personal control, task satisfaction, and task performance. Journal of Applied Psychology, 81(6), 738-745. doi: 10.1037//0021-9010.81.6.738.

81 Smith, M. (1992). Employee stress and health complaints in jobs with and without electronic performance monitoring. Applied Ergonomics, 23(1), 17-27. doi: 10.1016/0003-6870(92)90006-H.

82 Beaumont, P. (2011). Fresh claim over role the FBI played in suicide of Ernest Hemingway. The Guardian. Theguardian.com/books/2011/jul/03/fbi-and-ernest-hemingway

83 Anxiety and physical illness. (2012, July). Harvard Health Publications. Health.harvard.edu/newsletters/Harvard_Womens_Health_Watch/2008/July/Anxiety_and_phys ical_illness

84 Anxious people more likely to develop aggressive cancer, study finds. (2012, April 26). Metro. Metro.co.uk/2012/04/26/anxious-people-more-likely-to-develop-aggressive-cancer-study-finds-404181

85 Subašić, E., Reynolds, K. J., Turner, J. C., Veenstra, K. E., & Haslam, S. A. (2011). Leadership, power and the use of surveillance: Implications of shared social identity for leaders' capacity to influence. The Leadership Quarterly, 22(1), 170-181. doi: 10.1016/j.leaqua.2010.12.014

86 All Eyes On You. (2014). Psychologytoday.com/articles/201409/all-eyes-you

87 McDill, E. (1961). Anomie, Authoritarianism, Prejudice, and Socioeconomic Status: An Attempt at Clarification. Social Forces, 39(3), 239-245. doi:10.2307/2573215

88 McDill, E. (1961). Anomie, Authoritarianism, Prejudice, and Socioeconomic Status: An Attempt at Clarification. Social Forces, 39(3), 239-245. doi:10.2307/2573215

89 York, J. C. (2013, June 25). The chilling effects of surveillance. AlJazeera. Aljazeera.com/indepth/opinion/2013/06/201362574347243214.html

interest of authority figures deviates from the best interests of a society.[90],[91]

When Privacy Is Destroyed, Wealth & Liberty Are Destroyed. The gigantic corporations and governments that dominate our world today have repeatedly demonstrated that they cannot protect our privacy. They have allowed hackers to steal billions of sensitive citizen records, exposing a large majority of humanity to life-long identity theft and personal suffering, which would never occur if they didn't have our private data in the first place.[92] They have lobbied for and passed defective laws and regulations that perpetuate and amplify broken incentives within the global economy, which causes an endless series of market bubbles, severe wealth inequities, and privacy-killing Know-Your-Customer (KYC) laws.[93] Their belligerent and short-sighted foreign policies frequently blow up (figuratively and literally), resulting in an ever-expanding surveillance state and privacy-killing Anti-Money-Laundering (AML) laws to fight their so-called "War on Terror," which has wasted over $5.6 trillion of American wealth.[94],[95] All of their privacy-killing policies have made humanity *much less* financially and physically secure today and for generations to come.

Politicians Really Do Want to Control Your Mind. The top-secret MKULTRA Project was a decades-long CIA program between the 1950s and 1970s in which our own government experimented on American citizens and other foreign nationals, treating them like laboratory rats.[96] Among other inhumane goals, the primary purpose of MKULTRA was to develop *actual* mind-control technologies that would enable the CIA to completely control any human mind.[97] The explicit

90 Stanley Milgram Biography. (n.d.). Goodtherapy.org/famous-psychologists/stanley-milgram.html

91 Stanley Milgram Biography. (n.d.). Goodtherapy.org/famous-psychologists/stanley-milgram.html

92 Gemalto. (April 2018). Data Breach Statistics by Year, Industry, More. Breachlevelindex.com

93 Fifty Shades of Green: High Finance, Political Money, and the U.S. Congress. (2017, May 2). Roosevelt Institute. Rooseveltinstitute.org/fifty-shades-green/

94 Fifty Shades of Green: High Finance, Political Money, and the U.S. Congress. (2017, May 2). Roosevelt Institute. Rooseveltinstitute.org/fifty-shades-green/

95 Brown University. (n.d.). Costs of War on Terror. Watson.brown.edu/costsofwar

96 CIA v. Sims, 471 U.S. 159, 105 S. Ct. 1881, 85 L. Ed. 2d 173 (1985).

97 CIA OKs MK-ULTRA Mind-Control Tests. (2010, April 4). Wired.com/2010/04/0413mk-
Continues on next page.

targets of this sinister program included normal American citizens who opposed the Vietnam War, financially destitute humans who could be reprogrammed into robotic killing machines to assassinate domestic and foreign political leaders, and U.S. politicians whose minds could be controlled like zombies to pass legislation favorable to certain special interest groups.[98]

Politicians & Corporations Really Do Plot Against Citizens. To anybody who respects the U.S. Constitution and Bill of Rights, the Occupy Wall Street (OWS) movement in 2011 was perceived as a natural expression of constitutionally protected free speech and freedom of assembly. Anybody with eyes could see OWS was entirely peaceful. However, we know from subsequent FOIA lawsuits filed against the FBI that several companies on Wall Street and the FBI perceived the protesters very differently.[99,100] They classified the protesters as "terrorists," which triggers the escalated powers of the Patriot Act.[101] Specifically, the FBI and several companies on Wall Street perceived OWS as *such a grave existential threat* to their power that they colluded to "formulate a plan to kill the [OWS] leadership via suppressed sniper rifles" and they were "interested in developing a long-term plan to kill local Occupy leaders via sniper fire."[102,103] In fact, a Vice investigative report revealed:

> *[T]hroughout the FBI materials, there is repeated evidence of the FBI and Department of Homeland Security, American intelligence agencies really working as a private intelligence arm for corporations, for Wall Street, for the*

ultra-authorized

98 Given the behavior of some politicians, *maybe* the MKULTRA project is still active.

99 Gershman, J. (2014). FBI Ordered to Justify Shielding of Records Sought About Alleged 'Occupy' Sniper Plot. Blogs.wsj.com/law/2014/03/18/fbi-ordered-to-justify-shielding-of-records-sought-about-alleged-occupy-sniper-plot

100 The FBI vs. Occupy: Secret Docs Reveal "Counterterrorism" Monitoring of OWS from Its Earliest Days. (2012). Democracynow.org/2012/12/27/the_fbi_vs_occupy_secret_docs

101 How the USA PATRIOT Act redefines "Domestic Terrorism." (n.d.). ACLU. Aclu.org/other/how-usa-patriot-act-redefines-domestic-terrorism

102 September 17, D. S. P., & Lindorff, 2016 | Dave. (2016, September 17). Classic Who: FBI, Snipers & Occupy. Whowhatwhy.org/2016/09/17/classic-fbi-snipers-occupy

103 The FBI Is Hiding Details About An Alleged Occupy Houston Assassination Plot. (2014). News.vice.com/article/the-fbi-is-hiding-details-about-an-alleged-occupy-houston-assassination-plot

banks, for the very entities that people were rising up to protest against.[104]

Additionally, Mara Verheyden-Hilliard from the Partnership for Civil Justice Fund filed a FOIA lawsuit against the FBI to obtain over 100 pages of secret FBI documents produced during the OWS protests.[105,106] Ms. Verheyden-Hilliard said, "the documents say that they [FBI] know that this is a peaceful movement, that it is organized on a basis of nonviolence," yet, the USG invoked the "terrorism" threat to *lawfully* expand their power to violently suppress a peaceful citizen movement.[107,108]

Transforming Perfect Citizens into Terrorists and Misfits. The abuse of USG power during the OWS protests was bad, but that's just an obvious example. There exists an equally disturbing reality: Since the Patriot Act was spawned in 2001, now, at any moment, a rogue political official with mass surveillance power can obtain your private information and leak it to your boss, your spouse, the media, your religious community, or any place else to sabotage your life.[109] Even people *with nothing to hide* would be mortified by certain kinds of disclosures. The following scenarios are very common real-life events, which can be twisted to make anybody appear less than desirable. (The words in parentheses represent the hypothetical impact of disclosing each embarrassing piece of information to important people in your life and/or the general public.)

- **Web Browsing:** Reveals your *exotic media* browsing habits ("Bye-bye political career.") and frequent visits to WikiLeaks.

---

104 The FBI vs. Occupy: Secret Docs Reveal "Counterterrorism" Monitoring of OWS from Its Earliest Days. (n.d.). Democracynow.org/2012/12/27/the_fbi_vs_occupy_secret_docs
105 Ibid.
106 Reports - Big Brother America. (n.d.). Bigbrotheramerica.org/reports
107 Wolf, N. (2012, December 29). Revealed: how the FBI coordinated the crackdown on Occupy | Naomi Wolf. The Guardian. Theguardian.com/commentisfree/2012/dec/29/fbi-coordinated-crackdown-occupy
108 The FBI vs. Occupy: Secret Docs Reveal "Counterterrorism" Monitoring of OWS from Its Earliest Days. (2012). Democracynow.org/2012/12/27/the_fbi_vs_occupy_secret_docs
109 See the long list of constitutional violations that the Patriot Act has made "legal" at the ACLU's Patriot Act information page: Surveillance Under the USA/PATRIOT Act. (n.d.). Aclu.org/other/surveillance-under-usapatriot-act

("Are you a seditious traitor?")

- **Health Records:** Reveals your prodigious prescription for medicinal marijuana ("Sorry, I don't think you're qualified for this job.") and that cream for your embarrassing rash. ("Bye-bye social life.")

- **Online Shopping History:** Reveals your *eclectic* book reading list, including saucy romance novels ("Is your marriage in shambles?"), *The Anarchist's Cookbook* ("It was just a phase, I promise!"), *Mein Kampf* ("I was just curious . . . I don't want to join the Nazi Party, really!"), *The Mind of Jihad.* ("I'm just trying to understand why they hate us so much . . . no, I don't want to hijack a plane!")

- **Credit Card Charges:** Reveals your subscription to *Hacker Monthly.* ("No, I don't want to hack into any credit card databases!")

- **Bank Account Records:** Reveals the rich friend in Saudi Arabia who sent you a generous gift. ("It was a gift from my student exchange program . . . I've already told you: I . . am . . not . . a . . terrorist!")

- **Phone Records:** Reveals frequent calls to your Middle Eastern doctor who happens to have the same name as a radical jihadi who manages the website KillTheAmericanSatan.org. ("Who are the other members of your sleeper cell!?! . . . "I'm just a college student. I don't even know what a 'sleeper cell' *is*!")

- **Netflix History:** Reveals that you just watched "Catch Me if You Can" and "The Secret History of American Bank Robbers". ("Hmmm, *veerry* interesting.")

- **Email History:** Reveals at least 100 philosophically *interesting* emails with words like "terrorist," "freedom fighter," "NSA," "Prism," "Liberty," "Osama bin Laden," "president," "Ron Paul," "End the Fed," "assassinate," "bomb," "airplane," "encryption," "Jekyll Island," "gold standard," and hundreds of other *subversive* words. ("Put this suspect on the permanent airport customs Secondary Screening list.")

- **GPS History:** *Proves* that you were near the scene of a nearby bank robbery. ("Purely a coincidence with your Netflix history?

Sure, buddy.")

- **Voter Registration Records:** Reveals you voted for a political candidate who ran as an Independent. ("So, you want *that lunatic* to be president? You're not mentally fit to vote anymore, citizen.")
- **Surveillance & Sabotage:** Pictures of you entering a drug rehab center to visit *your friend* who is recovering from addiction are anonymously emailed to your boss. (Your boss to you: "It takes real courage to seek help—good for you!" . . . Your boss to his secretary: "Never let that loser into my office again. We can't have that kind of garbage in this company.")

Context is Everything. In their proper context, normal life events usually occur without a second thought. Unfortunately, when a society becomes dominated by a police state, all that benign information gets sucked into your ever-expanding *citizen profile*. Your profile can then be easily twisted and exploited by corrupt politicians to destroy your career and personal life if you ever question their public policies or threaten their power. To be clear, these are not merely hypothetical risks; the FBI and CIA have entire handbooks dedicated to secretly sabotaging citizens in everyday places that many people would never expect them to be.[110,111,112]

The Exorbitant Price of Justice. If a rogue government official decides to target you, average citizens have virtually no chance to defend themselves. "I'll take them to court!" you might say. Assuming your bank accounts have not already been frozen, thereby depriving you of food and an effective legal defense (a common tactic of the USG), the legal fees can empty your bank account so fast that most people can't afford the exorbitant price of justice. Without access to your money, you

---

110 Stockton, R. (2017, April 18). How The FBI Used Murder And Blackmail To Thwart The Civil Rights And Antiwar Movements. Allthatsinteresting.com/cointelpro-fbi

111 How the CIA's Fake Vaccination Campaign Endangers Us All. (n.d.). Doi.org/10.1038/scientificamerican0513-12

112 Secret CIA documents show the 9 best ways to annoy your boss. (2015, November 1). Independent.co.uk/news/world/americas/declassified-cia-documents-detail-how-to-sabotage-employers-annoy-bosses-a6716961.html

can't fight back. If you can't fight back, it becomes very difficult to survive a comprehensive character assassination and keep your career, your friends, and your family. Most people are forced to plead guilty or make deals that ruin their lives, which they could have avoided if they had enough money to properly defend themselves; or, if the rogue government didn't have all their private data in the first place.

The Convergence of A.I., Robots, IoT & Institutional Dysfunction. As this convergence and the deterioration of global trade and living conditions within and between nations continues to accelerate, we will see an escalation of class, ethnic, and international conflict in the years to come. As these tensions rise, the temptation for politicians to use high-tech tools to *keep the peace* will be irresistible. They will inevitably unleash these nonhuman, electronic creatures on humanity, deploying billions of tiny A.I. robotic spies against us, infiltrating every crevice of our lives. We know this is inevitable because it's the logical next phase of the expanding surveillance dragnet and concentration of power that already exists today.

Laws Don't Stop Politicians Who Think They're Above the Law. Some people think establishing secret FISA Courts and other purely administrative mechanisms is enough to prevent rogue politicians from abusing their power. It's not enough because every generation spawns new politicians who think they're above the law. For example, when asked whether President George H.W. Bush should have consulted the U.S. Congress before starting a war against Iraq in 1991, then-Secretary of Defense, Dick Cheney, said, "If we had lost the vote in the Congress, I would certainly have recommended that we go forward anyway."[113] Bush Sr. had enough sense and respect for the law to ignore Cheney's advice and Bush Sr. requested and received Congressional approval. However, Cheney lingered in the USG long enough to find somebody more pliable 10 years later—Bush Jr. . . . and the Patriot Act was born.

Preventing Unauthorized Access Is the Only Way to Prevent Government Abuse. Cheney's willful disregard for the Constitution is

---

113 Transcript | Cheney's Law | FRONTLINE | PBS. (2007).
Pbs.org/wgbh/pages/frontline/cheney/etc/script.html

just one example in a long history of over-zealous politicians who abuse their power. This is why it's so dangerous for any government to collect or have access to *any* private data on individual citizens *without probable cause and search warrants*, as explicitly stated in the 4th Amendment of the U.S. Constitution. At any moment, a rogue official can fabricate an excuse about why they can ignore the Constitution and violate your 4th Amendment rights. At any moment, a rogue official can search or seize your property, leak your intimate details to the press to publicly shame you for resisting their political agenda, or blackmail and intimidate you to prevent you from blowing the whistle on their illegal activities. The only way to prevent these abuses of power is to *prevent them from having that power in the first place* by using cryptographically secure communications and a cryptocurrency that truly protects your privacy and human rights. (Part 2 of this book is focused on that topic.)

Politicians & Corporations Often Conflate Self-Interest with the Public Interest. This is how spy programs like COINTELPRO, PRISM, Carnivore, XKeyscore, MYSTIC, OAKSTAR, and many other CIA and NSA programs get funded with taxpayer resources. Politicians and federal agents often perceive these programs as career-enhancing opportunities to make their personal mark on the world. They often have good intentions, but they either ignore or underestimate the consequences of their projects in the real world. Regardless of their intentions, these programs prove a basic truth: If we give politicians ubiquitous surveillance power, they will use it against any citizen that they think threatens their *personal interests*, which they frequently conflate with national interests.[114,115] That inevitable temptation represents a far greater threat to the long-term health and welfare of the general public than terrorists ever will.

Resistance Will Be Futile Long Before the A.I. Singularity. The day is coming when we will have no meaningful privacy from government surveillance, no meaningful institutional integrity in our economic and political systems, and no way to restore our privacy or

---

114 Timeline of NSA Domestic Spying 1791-2015. (2012, November 30). Eff.org/nsa-spying/timeline.
115 Churchill, W., & Wall, J. V. (1990). The COINTELPRO Papers: Documents from the FBI's Secret Wars Against Dissent in the United States. Boston, MA: South End Press.

institutional integrity because it will be too late. The technologies deployed against us will soon be functionally omnipotent and omniscient long before the dreaded A.I. *singularity* arrives. If we don't act soon to protect our privacy, our wealth, and our liberty, it will be too late . . . and resistance will be futile.

The Money Laundering Trap

The Vicious Cycle of State-Spawned Crime. The crimes that politicians claim to be trying to prevent with their ubiquitous mass surveillance apparatus are actually diseases caused by the deterioration of the economic and political systems that *they* are destroying. In fact, they are simply creating a series of self-fulfilling prophecies that will inevitably lead to all the disasters they claim to be trying to prevent. To understand how this vicious cycle works, let's briefly examine the four primary causes of money laundering, which has become an all-consuming obsession of USG officials since the early 1980s.

Money Laundering for Drug Trafficking. It's well-documented that Presidents Nixon and Reagan declared their illegitimate "War on Drugs" for political reasons, not because they cared about the health and welfare of American citizens.[116] The ongoing humanitarian and economic catastrophe Nixon and Reagan unleashed continues to plague large swaths of humanity to this day. A large portion of all global money-laundering is directly caused by this illegitimate war, which has become a war on liberty. Stop this illegitimate and immoral war and 25-50% of global money-laundering will disappear overnight. Thus, this source of money-laundering is not a legitimate reason to give up our human right to privacy and live like criminal rats in a panopticon cage.

Money Laundering for Terrorism Financing. Beginning with the 1953 plunder of the Iranian Government and the installation of the U.S.-backed puppet Shah (king); then the USG's alliance with Osama bin Laden in the early-1980s in Afghanistan to *defeat communism* when the USSR was already rapidly crumbling; then the USG's 66 covert global

---

116 See: Eanfar, F. (2017, August 22). The Secret History of the War on Drugs. Eanfar.org/secret-history-war-drugs

regime-change operations, nearly all of which resulted in humanitarian disasters; then Vietnam, Iraq and Afghanistan (*again*) . . . U.S. politicians have created a long list of enemies in the name of *spreading democracy*.[117] The human and economic cost of the collective hubris and ignorance of U.S. politicians has been devastating to millions of people who have died and millions more living today. Thus, when U.S. politicians antagonize, radicalize, and weaponize foreign nationals who are forced to resort to terrorism to fight the USG because they can't borrow USD trillions from China to build their own corrupt military-industrial-complexes, that's not a legitimate reason to give up our human right to privacy and live like criminal rats in a panopticon cage.

Money Laundering for Fraud. There are two general reasons that fraud proliferates in any society: (1) the institutional integrity of governments and corporations breaks down, which reduces accountability and teaches the rest of society to cheat because their political and business *leaders* are cheating; and (2) economic conditions deteriorate, which creates deprivation and existential desperation. When governments are competent and operate with institutional integrity, their societies are stable and prosperous, which substantially eliminates the incentives for citizens to engage in fraudulent activity.

In contrast, when politicians are corrupt and/or incompetent, systemic fraud proliferates throughout the society because large populations of humans are forced to break the law just to survive and compete with all the other *lawful* fraudsters that often control the largest banks, corporations, and governments. That means **widespread fraud is an institutional leadership problem**. Thus, the mere potential for fraudulent transactions in an economy is not a legitimate reason to give up our human right to privacy and live like criminal rats in a panopticon cage.

Money Laundering for Organized Crime. Anybody who has ever lived with a corrupt government and/or a dysfunctional economy knows that organized crime inevitably proliferates when economic conditions deteriorate. Most people have no need or desire to live the

---

117 See: Eanfar, F. We Are Paying the Price for Realpolitik. (2018, March 11). Eanfar.org/we-are-paying-the-price-for-realpolitik

dangerous life of a criminal when they have competent governments that know how to provide economic and social stability to their citizens. In fact, organized crime is a byproduct—*not a cause*—of *already-existing* government and economic dysfunction and corruption. That means policymakers should be spending all their time fixing the fundamental economic and institutional causes of organized crime, not forcing all citizens to give up their human right to privacy and live like criminal rats in a panopticon cage.

A Police State of Irrationality

Preventive Wars Are Like Committing Suicide to Prevent Death. Otto von Bismarck, one of the most revered statesmen and military strategists in world history, echoed this principle when he said, "Preventive war is like committing suicide for fear of death." In other words, unless a direct, specific, and clearly defined attack on the homeland is imminent within days or weeks, preemptively attacking another country to *avoid war* is like committing suicide to avoid death. Yet, this is exactly what many predatory U.S. politicians have said to justify destroying USD trillions of wealth and millions of human lives with their discretionary, offensive, *preventive* wars.

The War on Terror Violates Every Rational Cost-Benefit Analysis. If you asked any doctor or scientist, "How much taxpayer money should the government spend to prevent a disease that kills one person per year?" they would look at you like you're crazy. This is because approximately 611,000 Americans die of Heart Disease every year; 585,000 die of Cancer; 131,000 die of auto accidents; 129,000 die of stroke; 85,000 die of Alzheimer's Disease; 76,000 die of Diabetes. . . . Each of those diseases and many others inflict USD billions of damage to U.S. Citizens and the economy *every year*. Every death is tragic, but no rational cost-benefit analysis would conclude that $5.6 trillion of taxpayer funds should be spent to prevent one death per year. Yet, that's exactly how much U.S. politicians have spent on the *War on Terror* since 2001.[118,119]

---

118 Watson Institute. Brown University. (2018). Costs of War on Terror.
Continues on next page.

Is It Rational to Be Afraid of Terrorism? Each American has a 1 in 46,192,893 chance of being killed by a foreign terrorist. To put this minuscule risk into perspective, you are 265 times more likely to die from a bolt of lightning; 6.9 million times more likely to die of cancer; 82,000 times more likely to die in a car accident; 43,000 times more likely to die of a heart attack; 800 times more likely to be killed by a gun; 615 times more likely to be killed by an asteroid; and 366 times more likely to die randomly walking in a park.[120,121] If the risk of Americans dying from terrorism is so low, why have U.S. politicians spent over $5.6 trillion on the *War on Terror?*

American Taxpayers Have Paid More than $56 Billion *for each* Prevented Death Since 9/11. Millions of Americans suffer and die every year from thousands of preventable diseases and other events, which collectively cost the U.S. economy and taxpayers at least $1 trillion *every year* in healthcare expenses and lost economic productivity.[122] Yet, U.S. politicians have spent $5.6 trillion of our wealth to prevent *one statistical terrorism death per year* since 2001. In other words, *American taxpayers have paid approximately $56 billion per each death avoided since the 9/11 terrorist attacks.*

"It's Better to Be Safe than Sorry." This is a frequent claim made by supporters of the *War on Terror,* but this claim invokes the pretense of wisdom to conceal an absurdly illogical reality. The total property damage and economic cost of the 9/11 attacks was $50–120 billion.[123,124125,126] However, the property component of this cost was

---

Watson.brown.edu/costsofwar

119 Street 351 Pleasant, & MA, S.B. #442 N. (2018). Cost of National Security. Nationalpriorities.org/cost-of

120 Mueller, J., & Stewart, M. (2015). Chasing Ghosts: The Policing of Terrorism (1 edition). Oxford; New York: Oxford University Press.

121 The Economist, "Danger of Death! How you are unlikely to die" (February 14th, 2013), Online: Economist.com/blogs/graphicdetail/2013/02/daily-chart-7

122 "Milken Institute Study: Chronic Disease Costs U.S. Economy More Than $1 Trillion Annually"; Online: Fightchronicdisease.org/media-center/releases/milken-institute-study-chronic-disease-costs-us-economy-more-1-trillion-annual. Last accessed: October 12th, 2015.

123 Rose, Adam Z. and Blomberg, Brock S., "Total Economic Consequences of Terrorist Attacks: Insights from 9/11" (2010). Published Articles & Papers. Paper 190; Online: Research.create.usc.edu/published_papers/190

124 Mueller, J., & Stewart, M. G. (2011). Terror, Security, and Money: Balancing the Risks, Benefits, and Costs of Homeland Security (1 edition). Oxford ; NY: Oxford University Press.

closer to $50 billion and there is still substantial debate about the true economy-wide costs because U.S. GDP was already in free-fall prior to 9/11/2001 due to the Dot-Com bubble popping. Thus, many politicians in 2001 had political incentives to blame the economy's poor performance on 9/11, but GDP actually started growing *immediately after* 9/11. Although the psychological shock value was high, the *actual* nationwide economic impact of 9/11 was relatively low overall. Regardless, using a professional cost-benefit analysis to rationally justify the financial expenditures of the ongoing *War on Terror*, it would require a 9/11-scale attack on American soil *every week* to justify a $5.6 trillion (and counting) *War on Terror*.

No Threat on Earth Justifies Destroying Our Wealth, Privacy & Liberty. No terrorist organization *or country* has the capacity, skill *and incentive* to execute a 9/11-scale attack every week. Osama bin Laden himself never expected the World Trade Center buildings to collapse.[127] So, 9/11-scale attacks *were already* nearly impossible before 9/11. All that was required to make 9/11-scale attacks *more impossible* was to improve airport, seaport, and border security and certain FAA regulations, not a never-ending $5.6 trillion (and counting) *War on Terror*. Instead of acting based on a rational cost-benefit analysis and being honest with the American people, short-sighted U.S. politicians have murdered over 1 million humans, spawned ISIS and multiple refugee and humanitarian crises, and created a global surveillance dragnet that destroys the privacy, liberty, and wealth of billions of humans worldwide.[128] *The War on Terror is an ongoing crime against humanity.*

"American Society" *Does Not* **Support the** *War on Terror.* Several academic papers have been written about why *American society* spends so much money to prevent deaths from terrorism. Those

---

125 Bilmes, L. J., & Stiglitz, J. E. (2008). The Three Trillion Dollar War: The True Cost of the Iraq Conflict (52788th edition). New York: W. W. Norton & Company.

126 CNN (2013). September 11th Terror Attacks Fast Facts.
CNN.com/2013/07/27/us/september-11-anniversary-fast-facts/index.html

127 Harnden, T. (2001, December 10). Bin Laden didn't expect New York towers to fall. Telegraph.co.uk/news/worldnews/asia/afghanistan/1364905/Bin-Laden-didnt-expect-New-York-towers-to-fall.html

128 Watson Institute. Brown University. (2018). Costs of War on Terror.
Watson.brown.edu/costsofwar

academics often focus on psychological theories of the human response to terrorism fears, but those theories ignore a fundamental reality: "American society" *is not choosing* to spend USD trillions on foreign wars, enormous defense bureaucracies, and domestic spy programs. To the contrary, all major polls confirm that Americans feel *less safe* since 9/11; and they're much more afraid of small-scale attacks by home-grown sociopaths, not 9/11-scale terrorist attacks.[129,130] The threats that Americans are *really* concerned about are job security (an economic policy problem), domestic terrorists (a law enforcement, economic policy, and cultural problem), institutional corruption (a political leadership problem), and cultural deterioration (a political, economic, and corporate leadership problem). None of those problems justify a $5.6 trillion global *War on Terror.*

Who Really Supports the War on Terror? The *American people* as a whole have never supported the *War on Terror* or any of the obscene, wealth-, privacy-, and life-destroying policies associated with it. This undemocratic outcome is caused by a combination of: (1) professional incompetence in Congress and the White House; (2) self-serving politicians who are afraid of losing their jobs if an attack occurs during *their term;* and (3) deliberate deception intended to satisfy the special interest groups that profit from war and the global panopticon surveillance cage in which they're imprisoning all of humanity. In other words, it's only a tiny political cartel in Washington that willfully ignores the democratic will of the American people and consciously refuses to apply the most basic cost-benefit analysis to this grotesque *War on Terror.*

Do Americans Have a Better Place to Spend $5.6 trillion? There are many more productive ways to spend $5.6 trillion. For example, cures for disease (Heart Disease, Cancer, Diabetes, Alzheimer's, HIV...); develop renewable energy technologies, cold fusion, and safe Thorium-based nuclear power plants, which would completely eliminate electrical

---

129 Taylor, A. (2016, August 22). Poll: 42 percent of Americans say they are less safe from terrorism than before 9/11. Washington Post.
Washingtonpost.com/news/worldviews/wp/2016/08/22/poll-42-percent-of-americans-say-they-are-less-safe-from-terrorism-than-before-911
130 Director, J. A., CNN Polling. (2016). Poll: Concern about terrorist attack at highest level since 2003. CNN.com/2016/06/23/politics/terror-attack-poll/index.html

grid-based air pollution and dramatically reduce the world's dependence upon oil-producing authoritarian regimes, which would eliminate most sources of terrorist funding; upgrade the national infrastructure to increase nationwide productivity; harden the electrical grid to protect against natural and man-made EMP events; provide education scholarships and grants to more students; provide loans to small businesses to boost job growth; fix the broken healthcare system; improve millions more lives by allocating funds to other life-enhancing programs. . . . Or, just stop stealing wealth from the American people and let them spend their hard-earned money however they wish!

The U.S. is Losing the War on Terror and Much More. It is difficult for politicians to resist pressure from special interest groups to continue funding USD multi-trillion scams, but this is why leadership is so important. The U.S. Congress and White House are perpetually plagued by weak political leaders who make spending decisions based on how their next election will be affected, not the best interests of the entire nation. Until congressional and presidential candidates are elected on a nonpartisan campaign financing platform that eliminates their dependence on war-mongering special interests, the American people will continue to suffer from the ongoing grotesque theft of their wealth, unconstitutional destruction of their privacy and liberty, and a dramatic decline in the quality of their lives.

In Part 2 of this book, we will explore specific, concrete actions that the Gini Foundation is taking to resolve these and many other problems.

Key Points

- **Privacy Is a Fundamental Human Right.** All other rights depend on the right to privacy. Anybody who says privacy is not a human right does not understand the purpose of the Constitution or the 4th Amendment or the long history of manipulation and sabotage perpetrated by politicians and gigantic corporations against their citizens and customers. Democracy cannot exist without protecting the human right to privacy because politicians and the gigantic corporations that finance their political campaigns will inevitably use their wealth and power to spy, manipulate, and sabotage their political opponents. These assaults on democracy and human liberty are only possible when politicians and gigantic corporations are given the power to violate our human right to privacy.

- **Police States Make Their Countries Less Safe.** Police states spawn cultures that distrust authority, which increases the intensity and frequency of domestic terrorist attacks. Police states enable politicians to hide behind dubious secrecy laws that conceal their actions and decision-making processes, which inevitably leads to corruption, expropriation of wealth from citizens, jingoistic foreign policies that spawn foreign terrorism, all of which result in widespread tyranny and humanitarian atrocities. Police states create a constant state of anxiety throughout a population, which leads to widespread physical and mental health problems and the destruction of social cohesion within a society, ethnic and class tensions, conflict, violence, and a systemic breakdown of law, order, and peace.

- **The War on Terror is an Ongoing Crime Against Humanity.** The ongoing destruction of the Middle East—the cradle of human civilization—is a tragedy. The ongoing expropriation and destruction of $5.6 trillion (and counting) of American wealth is a tragedy. The ongoing destruction of human liberty by destroying our human right to privacy is a tragedy. The ongoing economic and cultural destruction of the American civilization by rogue politicians is a tragedy. Each of these ongoing tragedies is a crime against humanity. Collectively, they're a slow-burning human apocalypse.

- Chapter 3 -
An Epic Battle Is Coming

"To compel a man to furnish funds for the propagation
of ideas he disbelieves and abhors is sinful and tyrannical."
— Thomas Jefferson

Self-Interest vs. Humanity's Interest

Many people and organizations consciously or unconsciously confuse their self-interest with the broader interests of humanity. Sometimes people are acting in good faith and they truly don't see how their actions are hurting their communities or humanity as a whole, but often they do see it and they're too afraid to admit it because they don't want to lose their power and/or wealth. Regardless, this is a big problem in our world today because it corrupts political systems and economies, which frequently destroys our wealth and the lives of millions of humans. We see this problem frequently when, for example:

- Corporations export jobs to foreign countries without meaningful regard for the impact on their home economy; then they claim it's good for humanity.
- Corporations lobby for arbitrary import tariffs that protect *their industry*; then they claim it's good for humanity.[131]
- Corporations extract natural resources with poisonous chemicals and dump those chemicals into the environment; then they claim it's good for humanity.
- Corporations and puritanical libertarians claim that all taxes are

---

131 In *Broken Capitalism: This Is How We Fix It*, I provide a detailed analysis of the U.S. tariff system and ways to reform the tariff system worldwide to make our global economy more stable and equitable.

bad; or, that all government is bad; or, that all regulations are bad; or, that any restriction on their activities to create more ecosystem stability is an attack on humanity's liberty.

- Gigantic corporations and shareholders claim that their high concentration of market power creates economic efficiency that is in the best interest of humanity.

- Politicians claim that mass surveillance systems are in the best interest of humanity; when in fact, these systems are *primarily* (not exclusively) intended to protect their political careers and enrich the corporate executives and shareholders that finance their political elections.

- Politicians claim that their *War on Terror* and *War on Drugs* and the wars in Vietnam, Iraq, and dozens of other countries are in the best interest of humanity; when in fact, these wars are *primarily* (not exclusively) executed to enhance their political careers and enrich the corporate executives and shareholders that finance their political elections.

- Banks claim that bailing them out of bankruptcy after they destroy entire markets and economies is in the best interest of humanity.

- Central bankers claim that arbitrarily manipulating the money supply (often for political reasons to help certain politicians get reelected) is in the best interest of their citizens.

- Banks and politicians claim that consumption-based economies that emphasize debt over savings—thereby trapping billions of humans in a vicious cycle of life-long debt and dependence on banks and governments—are in the best interest of humanity.

- U.S. regulators and an army of lawyers, accountants, and *consultants* that feed at the regulatory trough often claim that the overwhelming blizzard of regulations that currently plagues the U.S. economy is in the best interest of all Americans.

- Fishermen, seal-hunters, whale-hunters, buffalo-hunters, timber producers, etc. sometimes claim that unrestricted harvesting and causing the extinction of those creatures is in the best interest of humanity.

- Among many other instances where special interest groups

claim that supporting their narrow agenda is in the best interest of humanity.

With an understanding of how politicians and corporations often confuse their self-interest with humanity's interest, we can avoid making the same mistakes when evaluating ideas and solutions for the major socioeconomic and geopolitical problems that humanity faces today.

The Epic Battle

The Epic Battle Over Tax Policy. At first glance, it might seem strange that a book about cryptocurrencies would discuss tax policy, but tax policy is the ultimate battleground where the epic fight for privacy and liberty will inevitably be won or lost in the future. The convergence of A.I., expanding corporatocracy, destruction of the middle class, never-ending wars, rapidly rising national debt, broken economic policies, and the dysfunctional and often corrupt federal budgeting and tax systems are going to make the tyranny much, much worse. The only way to prevent another bloody American civil war is to implement a politically viable, economically sustainable, and *privacy-friendly* alternative tax system that good-faith politicians and citizens can understand and embrace.

The Price of Government. Despite all the ideological noise and propaganda on both sides of the political spectrum, *government is a service* just like any other service and it should be priced according to the same law of supply and demand as other services. But this service *forces* citizens to sacrifice a portion of their wealth (taxes) to subsidize it; so, *demand for government service* must be based on a truly democratic process, not the arbitrary dictates of a tiny number of wealthy special interest groups. For example, in *a real democracy*, the demand and corresponding price of the service called "national defense" would certainly not be $600 billion per year; nor would a democratic society spend *$1.5 trillion for a single airplane program*; nor squander billions on obscenely wasteful government programs, bank bailouts, and multi-trillion-dollar national debts that steal—*yes, steal*—wealth from us every day.

What Services Should Government Deliver? Let's start with the *service categories* that a government definitely should *not* deliver: (1) multi-trillion-dollar *offensive* wars; (2) multi-trillion-dollar police-state infrastructures; (3) multi-trillion-dollar corporate welfare programs; (4) multi-billion-dollar *aid* to foreign dictators who brutally oppress their people; (5) multi-trillion-dollar bailouts and monetary system manipulation that erode the value of our money, destroy our wealth, destabilize the global economy, and discourage long-term savings; and (6) oppressive multi-billion-dollar federal bureaucracies to support all these wars, corporate welfare programs, oppressive foreign regimes, and a police state apparatus that violates our privacy and liberty every day.

Time = Money = Life. The income you earn represents a trade-off: your time in exchange for money. This is the essence of the *time=money* concept. Given that your life can only exist within the dimension of time, wasting your time equals wasting your life, which means we can also say "time=life." When any entity confiscates your money, that means they're confiscating all the time (which equals life) that was required to earn that money. When an entity confiscates your money, they are really confiscating equivalent portions of your life. Thus, time=money=life.

How Should a Government Spend a Tax That It Collects? Since time=money=life, this is the most important question of all. Partisan political operatives on both sides of the ideological spectrum like to cherry-pick facts to support various *big government* or *small government* agendas, but their ideological frenzy blinds them to a basic principle: The size of government is a *byproduct* of the spending priorities of the politicians who control the government. *Size* is not a policy objective; it's a policy *outcome*, a symptom of the collective fiscal priorities and ethical integrity of the creatures in control of the government.

The Size of Government Is Irrelevant When There Is Institutional Integrity. If a government's primary mandate is to deliver essential services with a genuine intent to create an economic and cultural environment that empowers citizens to enjoy the most personal freedom and the highest quality of life possible, then all public policies will naturally and efficiently flow from that mandate. In this context, it doesn't matter how big or small the government is as long as it is delivering the essential services necessary to create the highest quality of life for the largest number of citizens. This alignment of interests between the government and the citizenry is the essence of institutional integrity.

Mindless Propaganda About Social Welfare. Let's dismantle a fallacy that partisan operatives love to cram down the collective throat of the American people. They like to falsely accuse countries like Switzerland, Norway, Sweden, Finland, Denmark, and any other country with a strong social welfare system as "socialist," "borderline communist," "cradle-to-grave welfare states," and various other political slurs that reveal more about the ignorance of the speaker than the policies of the nations they attack. These people apparently don't understand how many trillions of dollars U.S. politicians have spent, and continue to spend, on the most atrocious crimes against humanity, which represent an incomprehensible waste of every American's time=money=life. These people mindlessly ridicule peaceful countries who spend their resources on providing the highest quality of life to their citizens instead of bombing, sabotaging, destroying, colonizing, economically and culturally raping and pillaging their own people and the entire planet. Do these people really think they are defending *democracy* or *free-market capitalism* or *American Exceptionalism?*

Taxes Are Not the Problem; How Politicians Destroy Our Wealth is the Problem. Some people like to say, "Total U.S. federal tax as a percentage of GDP is relatively low compared to many other OECD nations; so, what's the problem?" But they ignore the most important question of all: How do politicians actually spend your money? And yes, let's compare the performance of the USG to all the other OECD nations for all the following questions: Are taxes benefiting the broadest population of Americans? How much poverty is there in the United States? How many Americans live on the streets and in prisons? What is the median quality of life in the United States? How large is the median wealth gap between the poorest and richest Americans? What is the infant mortality rate? How are American students performing in primary and secondary school? How much and to what extent does systemic corruption impact policies? In what condition is the transportation and telecommunications infrastructure of the United States?

The USG is at the Bottom of All OECD Performance Rankings. The list of government performance questions above could have continued for several pages, but here's the point: By every meaningful measure of governmental performance and quality of life, *the United States ranks last or near the bottom among all OECD nations.*[132] And the reason is simple: U.S. politicians spend approximately 50% of all federal taxes on multi-trillion-dollar scams, which primarily benefit a tiny number of gigantic corporations and politically connected special interest groups, not the broader general public.

Illegitimate Governments Never Have Enough Money. It doesn't matter how high or low federal taxes are if the politicians who control the government are going to waste approximately 50% of the money anyway. Raise the taxes and that just gives them more money to waste. Lower the taxes and they will manipulate the money supply, destroy our purchasing power to subsidize their multi-trillion-dollar scams, find ways to take care of their corporate cronies, and cut the safety net for the humans that have the least power, which are usually the humans who genuinely need the most assistance.

3 Factors of National Stability: Rule of Law, Tax Revenue & Fiscal Legitimacy. Rule of law enables societies to enforce contracts; without this, economies don't work. Tax receipts enable governments to pay for essential government services; without this, civil society breaks down. A society's perception of the fiscal legitimacy of government expenditures determines the level of tax compliance and evasion, which determines the integrity of the preceding two factors (rule of law and tax receipts). Collectively, these three factors shape the incentives for humans and companies throughout an economy to trade their time, labor and capital for the net income that remains after all taxes have been paid.

How Much Tax Should We Pay? If taxes are too high or government expenditures are illegitimate, then there is little incentive for citizens to work. But if there's not enough taxation, then vital government services and infrastructure cannot be provided. So, taxes are obviously necessary; the primary questions are: *How* should we be taxed

---

132 Eanfar, F. (2018).Global Governance Scorecard.

and *how much* should we all pay? The answers are obvious, but the anti-democratic U.S. political system violates this obvious logic: The price of government (taxes) should only be as high as necessary to pay for the minimum level of government service *that a democratic majority of a nation's citizens want,* not based on backroom deals with disloyal corporate cannibals and deceptive legislative tricks.

Taxes Should be Fiscally Sustainable without Penalizing Productivity. A truly fair price of government service would be shared by all members of society, based on a fiscally sustainable methodology, without penalizing economically productive activity. Later we will explore an elegantly simple, equitable, fair, *nonpartisan,* and fiscally sustainable methodology. But before we focus on that, let's briefly discuss why the existing income tax system is illogical, immoral, unsustainable, and arguably unconstitutional.

Is the IRS Unconstitutional?

Depending on who you ask, the response to that question can range from "Are you a crackpot?" to "Hell, yes! It's an illegal extortion racket!" Many people have attempted to argue that the U.S. federal income tax (and the IRS in general) is unconstitutional, but they often lose their persuasive power when they get bogged down in arcane and subjective interpretations of various obscure court cases. We're not going to do that here. To the contrary, this will probably be the most interesting and intellectually stimulating tax policy discussion most people ever have. There are objective ways to analyze the tax system without resorting to court-case arcana and ideological sophistry. So, let's dive in.

The Gini Foundation's Perspective on Tax Policy Is Different. When we follow the money, we find that virtually all the tax policy *think-tanks* and nonprofit groups that promote various tax policies are funded by self-serving corporate special interest groups. More importantly, these groups are often trapped in self-serving, obsolete, and unsustainable notions about Economics that don't reflect the reality of how humans are affected by their policy prescriptions in the real world. As a result, their tax policy analysis and prescriptions are often inadequate for a modern world in which A.I., the expanding corporatocracy, broken incentives within the USG, and an ubiquitous surveillance state are all converging to destroy our privacy, liberty, and wealth.

Taxation Must be Based on Genuine Democracy. As we explore these fascinating topics, we should never forget why the American colonies declared their independence from Britain and why they fought the American Revolution: A democratic government's power to tax must be based on the consent of a majority of its citizens; otherwise, it is illegitimate taxation, which is, by definition, undemocratic tyranny.

The IRS Is a Political Weapon. This may shock some people, but the reality is that virtually every U.S. presidential administration since the IRS' inception in 1913 has used the IRS as a political weapon against people and organizations that challenged their power.[133,134,135] The IRS has compiled the most comprehensive database of citizen and organizational profiles on Earth. This all-seeing agency is not a *potential* threat; it's an *actual threat* to anybody who wants to hold U.S. politicians accountable for their performance or question their legislative priorities.[136,137,138] In fact, corrupt U.S. politicians and federal agents have repeatedly proven that they will not hesitate to "use the available federal machinery to screw our political enemies," according to former White House legal counsel John Dean.[139,140]

The Global Diaspora Tax is the Evil Root of the U.S. Police

133 The IRS's long history of scandal. (2013, June 8). Theweek.com/articles/463448/irss-long-history-scandal

134 Burnham, D. (1990). A Law Unto Itself: Power, Politics, and the IRS. NY: Random House.

135 III, J. A. A. (2002). Power to Destroy: The Political Uses of the IRS from Kennedy to Nixon (1st edition). Chicago: Ivan R. Dee.

136 IRS's rich history of scandals, political abuse. (2013, May 17). Reuters. Reuters.com/article/us-usa-tax-irs-scandals/factbox-irss-rich-history-of-scandals-political-abuse-idUSBRE94F16V20130517

137 Pilkington, E. (2014, January 7). Burglars in 1971 FBI office break-in come forward after 43 years. The Guardian. Theguardian.com/world/2014/jan/07/fbi-office-break-in-1971-come-forward-documents

138 Burnham, D. (1989, September 3). Misuse of the IRS: The Abuse of Power. The New York Times. Nytimes.com/1989/09/03/magazine/misuse-of-the-irs-the-abuse-of-power.html

139 The Sordid History of IRS Political Abuse. (2013, September 1) Fff.org/explore-freedom/article/the-sordid-history-of-

140 Nixon's Enemies List. (2018, March 22): https://en.wikipedia.org/wiki/Nixon's Enemies List

State. The USG's global "Diaspora Tax" enables U.S. politicians to directly or indirectly control every aspect of a U.S. Citizen's physical and economic life no matter where they live on Earth. This prevents Americans from voting with their feet and escaping the USG's corrupt debt prison. If you're not involved in any personal or business activities beyond U.S. borders, then you may not realize how overwhelming it is to deal with this constant bureaucratic barrage of tax regulations. These include "FBAR," "Form 8938," "Form 3520," "Form 3520-A," "Form 5472," "Form 926," "Form 8865," "FinCEN," "FATCA," "AML/BSA" and many other tax laws and regulations.

The IRS Causes Americans to be Rejected by Other Countries. If you're not familiar with most of those IRS forms and acronyms, don't worry, they all amount to one simple reality: If you're a U.S. Citizen or permanent U.S. resident, you can no longer effectively do business or live anywhere on Earth without paying the government for every breath you take, literally. And because U.S. politicians use punitive withholding taxes to bully foreign banks into complying with its oppressive taxes and its immoral *War on Terror*, the vast majority of foreign banks won't serve individual or corporate American clients anymore.

The Diaspora Tax *Is* Taxation without Representation. "Widespread enslavement," "imprisonment," "torture," "murder," "surveillance and censorship to create a culture of permanent fear"—these are the words used in a 2011 United Nations report that described an authoritarian country imposing a "diaspora tax," which "may amount to crimes against humanity" because it has attempted to "collect [2%] taxes outside of Eritrea from its nationals. . . ." That U.N. report was focused on Eritrea's tiny 2% diaspora tax, but Eritrea's human rights violation should sound familiar to all Americans: It's called "taxation without representation."

The Diaspora Tax Is a Human Rights Violation. In fact, it's such a serious violation of human rights that it inspired the U.S. Declaration of Independence, the American Revolution, and U.N. Security Council Resolution 2023, which demands that Eritrea "cease using extortion, threats of violence, fraud and other illicit means to collect taxes outside of Eritrea from its nationals or other individuals of Eritrean descent." In this case, Eritrea's crime against humanity is tame compared to what U.S. politicians do to their citizens. To understand why this is true, simply replace every instance of "Eritrea" above with "U.S. Government" and then increase the diaspora tax to *a minimum* of 50% and you will understand why this USG human rights violation is at least 25 times more despicable than Eritrea's 2% diaspora tax.[141]

What's Wrong with Diaspora Taxes? Because time=money=life, taxes are not just money coming out of your pocket; taxes are how corrupt and dysfunctional governments literally steal your life. Each minute, dollar, and quantum of life-force stolen by profligate politicians represents a minute, dollar, and quantum of life-force that you no longer have to spend with your family and friends. Since each human life is so short, this is a very serious human rights violation. This is why it's a "crime against humanity." Of course, the USG is the largest financial donor to the U.N.; so, the U.N. does not have enough political independence to explicitly condemn the USG with a U.N. resolution "to cease using extortion, threats of violence, fraud and other illicit means to collect taxes outside of the United States from its nationals or other individuals of American descent," but that's certainly what a financially and politically independent U.N. would do.

The Diaspora Tax Blocks Political Accountability and Reform. By taxing U.S. Citizens for breathing anywhere on the planet, the Diaspora Tax produces the following negative outcomes:

(1) Silences and discourages peaceful political dissent by granting the IRS *lawful authority* to unconstitutionally freeze your bank accounts and incarcerate you for your constitutionally protected free speech, while *officially* claiming you are being punished as a *criminal tax evader.*
(2) Blocks meaningful political reform because the label "criminal tax evader" destroys your professional reputation, which substantially prevents you from successfully building any new enterprise that could financially support a meaningful political

---

141 See also IRS Internal Revenue Manual ("IRM") §4.26.16.6.8, in which the IRS has a mandatory statutory obligation to confiscate a minimum of 50% of all undisclosed offshore accounts owned by U.S. Citizens if they collectively exceed $10,000. But it gets worse, the IRS is authorized by Congress to confiscate 100% of all undisclosed accounts, at the IRS examiner's discretion, regardless of where the citizen lives anywhere on the planet.

reform campaign.

(3) Transforms the law enforcement apparatus of the entire federal government into a weapon that politicians can use to evade personal accountability for their fraud and abuse.

(4) Gives politicians diplomatic pressure tools to bully foreign governments into extraditing you back to the U.S. if you try to organize a reform campaign while living anywhere else on the planet.

(5) Perpetuates the immoral multi-trillion-dollar extraction of American wealth to pay for endless wars, domestic and foreign political and economic oppression, corporate welfare programs, and an ever-expanding federal bureaucracy.

(6) Creates a constant distraction and drain on your time and personal resources, which can erode the passion and energy that liberty-minded Americans would otherwise have to continue fighting for their political and economic freedom.

Americans Have Two Bad Choices: Renunciation or Resignation. The Diaspora Tax forces U.S. Citizens to make an agonizing choice: (1) give up your U.S. citizenship through a tedious, humiliating, and expensive legal renunciation process; or (2) give up your economic and political freedom through a painful and humiliating process of lifelong resignation. Renouncing your citizenship means you have no ability to advocate for U.S. political reform within the borders of the United States, which substantially silences you. Giving up your political and economic freedom by resigning yourself to a life of quiet desperation also renders you impotent to advocate for any meaningful political reform, which substantially silences you. It's no accident that both of these bad choices lead to the same outcome: you are silenced and neutralized.

IRS Income Tax Forms & Audits Violate the 4th Amendment. The 4th Amendment states:

The right of the people to be secure in their persons, houses, papers, and effects, against unreasonable searches and seizures, shall not be violated, and no Warrants shall issue, but upon probable cause, supported by Oath or affirmation, and particularly describing the place to be searched, and the persons or things to be seized.[142]

In violation of the clear and explicit language of the 4th Amendment, income tax forms are an annual interrogation of the most intimate affairs of every citizen's life. (Of course, IRS audits are much worse.) And contrary to the popular myth of the "1-page income tax form in 1913," IRS income tax forms have been tedious, intrusive, multi-page interrogations since the IRS' inception in 1913.[143] In fact, *tax season* is an enormously frustrating, expensive, and time-consuming experience. Many people have been indoctrinated into believing this is normal, but it's not. The explicit language of the 4th Amendment was intended to protect citizens from any form of unnecessary search, seizure, and infringement on their privacy. All the time, money, and intimate personal disclosures required to complete an income tax return obviously violates the letter, intent, and spirit of the 4th Amendment.

The U.S. Tax System Violates the 8th Amendment. The 8th Amendment states the government shall not impose "cruel and unusual punishments." Yet, some people literally break down into tears because of how daunting, tedious, time-consuming, and disruptive *tax season* is for them. Many people and small companies can't afford to pay somebody else to do their taxes; so, *tax season* is *tax hell* for them. When a government imposes a burden that is so disruptive, costly, and painful that otherwise stable humans break down into tears, that's a big problem. Thus, it's not unreasonable to describe *tax season* as "cruel and unusual punishment" for being an American citizen. Regardless of whether the Supreme Court accepts this argument or not, there is no question that the U.S. tax system is an *actual* form of tyranny.

The IRS Costs Taxpayers $1 Trillion Annually. "Americans face up to nearly $1 trillion annually in hidden tax-compliance costs, while the Treasury forgoes approximately $450 billion per year in unreported taxes," according to the nonpartisan Mercatus Center at George Mason

---

142 Strasser, M. R. (2008, July 16). Fourth Amendment. Law.cornell.edu/wex/fourth_amendment
143 Tax History Project — U.S. 1040 Tax Forms, 1913 to 2006. (n.d.).
Taxhistory.org/www/website.nsf/Web/1040TaxForms

University.[144] Those costs include the time and money spent by Americans preparing tax forms, special interest lobbying expenditures, the federal budget to run the IRS' operations, and lost economic growth caused by *tax season*, which stifles productive activity throughout the entire economy. The $450 billion in "unreported taxes" is caused by tax evasion, which would be much lower if more Americans believed their taxes were spent on legitimate expenditures.

The Cost of the U.S. Tax System = 40% of All Taxes Collected. In 2012, when the Mercatus study was conducted, the IRS collected $2.5 trillion in total tax receipts. The total cost to the U.S. economy to collect those taxes was approximately $1 trillion. Thus, the annual cost to American taxpayers *just to collect our taxes* within the current broken tax system is up to 40% of all tax receipts collected. That's more wealth stolen from Americans every year because of the systemically corrupt and dysfunctional federal tax system.

The U.S. Tax System Violates the 5th Amendment. The 5th Amendment states that citizens shall not "be deprived of life, liberty, or property, without due process of law; nor shall private property be taken for public use, without just compensation." There is no doubt that the U.S. tax system destroys and steals our wealth through multiple mechanisms. Yet, some people still claim that "due process of law" occurs because the U.S. Tax Code is based on *tax laws*; thus, there's no 5th Amendment violation. This is like saying, "politicians have passed laws to destroy and steal your wealth; so, it's legal. Have a nice day!" When a legal system cannot protect the wealth of its citizens from the avarice and incompetence of politicians, and when the legal system codifies into law the wholesale destruction and theft of our wealth, the *legal system* is an undemocratic tyranny.[145]

The U.S. Tax System Violates the 13th Amendment. The 13th Amendment states: "Neither slavery nor involuntary servitude, except as a punishment for crime whereof the party shall have been duly convicted, shall exist within the United States. . . ." The USD trillions that U.S. politicians destroy and waste and the rapidly rising $22 trillion

---

144 Mercatus Center, George Mason University. (2013, May 20). The Hidden Costs of Tax Compliance. Mercatus.org/publication/hidden-costs-tax-compliance

national debt can *never* be paid back.[146] The vast majority of Americans did not vote for that odious debt, nor did they benefit from it. The national debt has trapped all living and unborn Americans in a life-long debtor's prison. That prison is built upon the unjust and undemocratic U.S. tax system, which politicians use to fund all their unjust wars and illegitimate corporate welfare scams. Clearly, U.S. politicians have condemned all Americans to a state of involuntary servitude, which is a government tyranny that has no regard for the democratic will of its citizens.

The 16th Amendment Ignores the Purpose & Intent of the U.S. Founders. The 16th Amendment states:

> *The Congress shall have power to lay and collect taxes on incomes, from whatever source derived, without apportionment among the several States, and without regard to any census or enumeration.*

There are *many* arguments that have been presented to the IRS and Supreme Court regarding the 16th Amendment, which was passed in 1913. Most of these arguments are based on very nuanced, subjective, and often unsubstantiated technicalities. So, the USG has predictably rejected all of them. However, the argument that is unassailable is this: The U.S. Constitution that the U.S. Founders created absolutely and explicitly prohibited *un-apportioned direct taxation*, which includes direct personal income taxation. This can be easily confirmed by the plain text of the Constitution itself:

> **Article I, Section 2, Clause 3:** *Representatives and <u>direct taxes shall be apportioned</u> among the several States which may be included within this Union, according to their respective Numbers. . . .*[147]
>
> **Article I, Section 8, Clause 1:** *The Congress shall have Power to lay and collect Taxes, Duties, Imposts and Excises, to pay the Debts and provide for the*

---

145 There are other problems in the U.S. legal system, but they're beyond the scope of this book.

146 Visit UsDebtClock.org to see the absurd condition of the U.S. economy.

147 Emphasis added. Article 1, Section 2, Clause 3 was intended to prevent unjust taxation in less populous states and to ensure that more populous states contribute a larger portion of total federal tax receipts.

common Defence and general Welfare of the United States; but <u>all Duties, Imposts and Excises shall be uniform</u> throughout the United States.[148]

Article I, Section 9, Clause 4: *No Capitation, or other direct, Tax shall be laid, <u>unless in proportion</u> to the Census or Enumeration herein before directed to be taken.*[149]

What is "Apportionment"? Until 1913, the U.S. Government was funded primarily by customs duties (tariffs) and excise taxes on *sin products* like alcohol and tobacco.[150] This is because the Constitution's Apportionment Clause required all taxes to be either <u>indirect</u> *and* "uniform" (e.g., a uniform tax of 3% on all tobacco sales *nationwide*) *or* <u>direct</u> *and* "apportioned" according to how many representatives each state had in Congress. For example, total property taxes paid by all the residents of Montana needed to be proportional to Montana's share of the total U.S. population, but the Montana legislature could determine how to divide the state's total share of property taxes according to a formula approved by Montana residents. The apportionment requirement was intended as way to account for how many people each state has, and thus, estimate how much of the public infrastructure and federal government's services each state consumes. In other words, states that consume more federal services should pay a larger portion of total federal expenses and taxes.

The First Federal Income Tax Was Born in the Civil War. If you understand everything up to this point, then you understand exactly what the U.S. Founders intended for the U.S. tax system. Now, what happened *after* the Constitution was originally written is where all the debate and confusion starts. The major debates started around 1861 when President Abraham Lincoln wanted to impose a 3% flat income tax to pay for the Civil War, which resulted in the passage of the Revenue Acts of 1861–64. However, the Revenue Acts were repealed after the Civil War and the federal government reverted back to funding

---

148 Emphasis added. Article 1, Section 8, Clause 1 was intended to prevent prejudicial federal taxation against individual taxpayers.

149 Emphasis added. Similar purpose as Section 2, Clause 3 above, but Section 9, Clause 4 adds the Census mechanism to accurately enforce Clause 3.

150 Buenker, John D. (1981). "The Ratification of the Sixteenth Amendment." The Cato Journal.

its operations with tariffs and sin taxes until 1913.

Who Wanted an Income Tax? A coalition of political parties, including the Socialist Labor Party, the Populist Party, the Democrat Party, among others wanted to change the tariff-based tax system to an income-based tax system for two primary reasons: (1) tariffs directly increase the price of commodities, food, and basic necessities, which impacted the poor and middle class throughout the country much more than the wealthier people in the Northeastern states; and (2) they were alarmed at the high concentration of financial power during the Gilded Age, which enabled people like John D. Rockefeller, J.P. Morgan, Andrew Carnegie, and others to use their financial power to control the political system and block economic reforms that the poorer constituencies needed.[151]

The 16th Amendment Was a Revolt Against Power Concentration. After the Civil War, several Supreme Court cases failed to resolve the rising tensions between the rich and poor throughout the country. So, with broad popular support throughout the South, Midwest, and Southwestern states, the 16th Amendment was passed in 1913. This created a permanent, progressive U.S. income tax. It's important to understand that this was not a conspiracy by the federal government or a socialist revolution; it was simply the result of decades of persistent democratic activism by a large portion of the American population, which was reasonably concerned about the rapid concentration of economic and political power in the Northeastern states. The obscene concentrations of wealth during the Gilded Age and the corresponding concentration of political power and corruption *inevitably* ignited a major populist uprising. The 16th Amendment was the result.

Is an Income Tax Economically Logical? A government can choose between many kinds of taxes, but one of the most illogical taxes is the income tax. There is no economically rational way to justify taxing the productive output of a society. Taxing anything means you will get less of it; so, taxing the income from productive output means a country

---

151 Palmer, Bruce (1980). "Man Over Money": the Southern Populist Critique of American Capitalism. Chapel Hill: University of North Carolina.

will get a less productive society. The fact that politicians ignore this basic economic principle is another glaring example of how undemocratic the political system has become. A truly democratic government would employ a tax system managed by people who understand basic economics, not politicians who use the tax system as a political weapon and a system of social control.

It's Time for Tax System 3.0. The U.S. *Tax System 1.0* lasted 126 years from 1787 to 1913. The *2.0* version has lasted for over 105 years from 1913 to 2018. Today, there are about *175,000 pages* of federal laws and regulations; approximately 74,000 pages are devoted to the federal tax code alone.[152] No human has ever read that entire tax code monstrosity. A rational person would say, "Why is the tax code so complex? Why don't politicians simplify it so that there are not so many loopholes and so much pain for American citizens to endure every tax season?" There are many reasons, but they all boil down to this: political inertia, political corruption, and a high concentration of financial power controlling the U.S. Government. Those are the same toxic conditions that spawned the 16th Amendment in 1913.

Is the U.S. Income Tax Unconstitutional? Using the IRS as a political weapon is tyranny. Forcing Americans to waste nearly $1 trillion per year on tax compliance costs alone is tyranny. Creating endless loopholes that give gigantic corporations embedded systemic advantages is tyranny. Criminalizing an ever-expanding portion of human existence by trapping Americans in an ever-expanding planetary web of mind-numbing tax laws is tyranny. Condemning all living and unborn Americans wherever they live on Earth to involuntary slavery is tyranny. Using the IRS as a weapon to steal our wealth is tyranny. Taxing Americans even when they don't live in the U.S. is taxation without representation, which is tyranny. Forcing Americans to make the agonizing choice between renunciation and resignation is tyranny. A legal system that defends a broken, antiquated, and systemically corrupt tax system with a blizzard of tax laws and arcane case law is tyranny. Regardless of its intended purpose, the aggregate impact of the current U.S. tax system results in unconstitutional outcomes; and 105 years of

---

152 Ten Thousand Commandments (2015). Competitive Enterprise Institute. Cei.org/10kc2015

this tyranny is enough. It's time for *Tax System 3.0*.

Reviewing Our Purpose

What Is Our Goal Here? To be clear, we are not suggesting that any citizen should break any law. We are not *tax protesters* trying to avoid paying *lawfully* imposed taxes. We are *lawfully* discussing important socioeconomic and institutional problems that are destroying the time=money=life of millions of humans in the U.S. and worldwide. By now, it should be clear that there is overwhelming, verifiable evidence to support the claim that the U.S. tax system is grotesquely wasteful, unjust, undemocratic, abusive, and egregiously injurious to American citizens, and indirectly devastating to millions of humans worldwide who suffer from U.S. taxpayer-funded wars and aggression.

A Web of Well-Intended Laws Can Produce Unlawful Outcomes. A single leech is harmless to a human, but a bucket full of leeches can suck all your blood in five minutes. Tiny things in large quantities can be deadly. In this context, whether or not any particular Constitutional provision has been technically violated is largely irrelevant to the actual real-world outcomes of the U.S. tax system today. Any particular law, when examined in isolation, can appear to be technically *lawful*, but a web of laws in aggregate can create such a stifling and destructive impact on a population that their collective real-world effect produces an emergent tyranny that undermines the spirit and intent of lawmakers and destroys the well-being of citizens.

A Crime Against Humanity. What we are demonstrating in this chapter so far is that any tax system on Earth that substantially destroys and/or misappropriates the wealth of millions of citizens is manifestly defective, immoral, and fundamentally *a crime against humanity*. That statement might surprise some people, but a "crime against humanity" has a specific legal definition, which the United Nations defines as:

[A] 'crime against humanity' means any of the following acts when committed as part of a widespread or systematic attack directed against any civilian population, with knowledge of the attack:

- *Murder;*
- *Extermination;*
- **Enslavement;**
- *Deportation or forcible transfer of population;*
- **Imprisonment or other severe deprivation . . .;**
- **Torture;**
- *Rape, sexual slavery . . .;*
- *Persecution against any identifiable group . . .;*
- *Enforced disappearance of persons;*
- *The crime of apartheid;*
- **Other inhumane acts** *of a similar character intentionally causing great suffering, or serious injury to body or to mental or physical health.*[153]

Tyranny's Legal Strategy. Lawyers defending the IRS would likely try to claim that the acts of "Enslavement," "Imprisonment or other severe deprivation," "Torture," and "Other inhumane acts . . ." do not apply to tax systems for various reasons. Such arguments are a matter of interpretation, intent, and the technical legal definition of each of those acts and whether the actual outcome of the tax system does in fact produce those outcomes. Obviously, that discussion would become another legal circus, which gives the IRS the home-field advantage. They will just continue invoking the self-referential and circular logic of a web of historical case law to bog down the litigation for years. That's not a winning strategy for the citizenry.

From the IRS' Perspective, All Tax Arguments Are "Frivolous". The IRS and Supreme Court fight every case with an endless spiderweb of circular logic, which never actually addresses the fundamental tyranny of a tax system that is used as a weapon, as a debt slavery trap, and as a mechanism of wealth destruction and theft. All the cases focus on extremely narrow questions that always have the same result: The power of the government to tax is absolute and all arguments

---

153 United Nations Office on Genocide Prevention and the Responsibility to Protect. (n.d.). UN.org/en/genocideprevention/crimes-against-humanity.html
Bolded emphasis added to the bullet points that are potentially applicable to our discussion here.
Some bullets were truncated due to spatial constraints.

to the contrary are "frivolous." Then, if the citizen believes the verdict is unjust, they can be imprisoned and spanked with very harsh penalties for promoting or pursuing "frivolous tax arguments." [154]

What Have We Learned So Far? Under the laws of the United States of America today, if you're a U.S. Citizen or permanent U.S. resident, you can be . . .

- hunted relentlessly;
- taxed into poverty;
- stripped of your political and economic liberty;
- forced at gun-point to comply with a human rights-violating diaspora tax that the United Nations classifies as a "*crime against humanity*";
- extradited back to the U.S. if you try to escape this tyranny;
- imprisoned for years if you try to live outside the United States without paying tribute to the USG just for breathing;

. . . and the only other government that behaves like this is Eritrea, a despotic and notorious violator of human rights in Africa.

Obviously, Politicians Will Never Give Up Their Taxing Power. There's no point in trying to fight this tyranny on the IRS' own terms because, no matter what any citizen says, politicians and legions of bureaucrats and corporate lobbyists that benefit from the status quo are never going to admit that the existing system is unconstitutional because that would prevent the federal government from funding itself. So, the most realistic path to justice is to present an alternative tax methodology to the general public, explain to citizens and politicians how it will generate equal or greater tax revenue for the government than the status quo, work to generate grassroots support for it that politicians can't ignore, then citizens should not vote for any politician that doesn't support the more rational tax system. That is our purpose with respect to U.S. tax policy.

---

154 See the IRS' Publication: "The Truth About Frivolous Tax Arguments" if you want to see how the IRS deals with all the most common tax protestor arguments. IRS.gov/pub/irs-utl/friv_tax.pdf

In Part 2 of this book, we present a politically viable and economically sustainable alternative to the existing *tax monster*, in addition to solutions for several other important socioeconomic problems.

Key Points

- **The U.S. Tax System Produces Unconstitutional Outcomes.** Regardless of the intended purpose or *lawfulness* of any particular tax law, the aggregate web of U.S. tax laws is stealing massive amounts of time=money=life from all Americans. This certainly produces tyrannical and unconstitutional outcomes.

- **The IRS Is a Blight on the American Spirit.** The IRS has become an impediment to American progress. Americans are plagued by the bureaucratic inertia, political sclerosis, and confiscatory tyranny of the IRS. This tyranny systematically destroys the value creation process and suffocates economic productivity, industrial innovation, global economic competitiveness, and meaningful political reform, which perpetuates many class, ethnic and social divisions throughout American society.

- **Tax Policy Is the Ultimate Battleground for Privacy & Human Rights.** As U.S. politicians continue to accumulate incomprehensible, anti-democratic, and odious debts, they are going to use the tax system to engage in ever-more oppressive forms of taxation and tax enforcement to pay for their systemically corrupt and short-sighted fiscal and monetary policies. As ever-more Americans become consciously aware of the time=money=life principle, they're not going to tolerate the status quo. No large population will tolerate economic and political tyranny forever.

Part II
Building a Better World

- Chapter 4 -
Meet the Real Adam Smith

"No society can surely be flourishing and happy,
of which the greater part of the members are poor and miserable."
— Adam Smith

Welcome to Part 2. This part of the book is focused on solutions. Of course, we still need to discuss solutions in their proper context to make sure we understand how and why the solutions apply to specific problems, but the solutions will be the primary focus from this point forward. One of the contextual factors that we will discuss in this short chapter is: What economic *philosophy* should we have if we want to fix all the problems described in Part 1? Capitalism? Communism? Socialism? Marxism? Libertarianism? Keynesianism? Neoliberalism? Do we need to conform to any *ism* at all; or, can we define a new economic philosophy that is economically and politically viable?

What Did Adam Smith Believe? All major economic philosophies today—including Capitalism, Socialism, Marxism and Communism—owe a significant intellectual debt to Adam Smith. Most people have never read Smith's book, *The Wealth of Nations*, which exceeds 1,100 pages. This is understandable considering the 100s of pages that Smith devotes to mundane topics like exchange rates, corn production, purchasing power parity, labor wage rates, the virtues of the division of labor, and folksy business management advice. However, without an accurate understanding of Smith's perspective on capitalism, corporate behavior, government intervention, taxation, and social welfare, it's impossible to verify whether economic and trade policies today are consistent with Smith's widely respected, nonpartisan theories on capitalism and free-market economics. Thus, it's useful for us to review what Adam Smith actually believed *in his own words*.

The Definition of a "Smithian Economist". As Enlightenment

Age humanists, economists following in the tradition of Adam Smith have historically taken a *holistic* approach to economic analysis. In other words, they integrate all knowable domestic and foreign factors into their analysis and they base their observations and conclusions on *real-world* phenomena and corresponding *real-world data*, not fictionalized, abstract models and rabid, self-serving ideology. Because the term "holistic" relates to the philosophical and methodological characteristics of an economist's analytical approach, there can be some occasional overlap between *classical* liberal and so-called *neoclassical* liberal economists who may use model-driven analysis in some contexts, but not in others. To keep our classification process as concrete and simple as possible, and to avoid confusion later, let's state the following:

> When an economist integrates all reasonably knowable and relevant domestic and foreign factors into their analysis and they base their observations, conclusions, and economic policy prescriptions *primarily* on *real-world phenomena* and corresponding *real-world data*, they are classical liberals, aka, "Smithian economists." In contrast, when they base their observations, conclusions, and economic policy prescriptions *primarily* on fictionalized, abstract mathematical models and/or self-serving ideology that artificially ignore important real-world factors, they are "neoliberal economists."[155]

The Qualified Virtue of Maximum Production. Both Smithian *and* neoliberal economists believe human welfare can be improved by encouraging corporations and nations to maximize their production and profit. However, the influence of Enlightenment Age humanism on *Smithian* economists typically results in them giving much higher priority to measures of *sub-national* human welfare in their economic theories and models, including their *distributional consequences*.

---

155 "Neoclassical Economics" is associated with the body of technical economic theories that emerged early in the 20th Century; whereas, "neoliberalism" is a more general ideology that has dominated global trade and economic development policies since the 1980s. For a succinct overview of the difference between these two philosophies and the crucial distinction between political liberalism and economic liberalism, see: Economic Liberalism vs. Political Liberalism. Eanfar.org/economic-liberalism-vs-political-liberalism
See also: Classical Liberalism vs. Neoliberalism. Eanfar.org/classical-liberalism-vs-neoliberalism

What Did Adam Smith Really Say?

Adam Smith on the Behavior of Capitalists & Corporations. Smith obviously appreciated the concepts of capital allocation, free markets, the division of labor, prudential risk management, supply and demand, and all the ideas commonly associated with capitalism, but what did he think about *capitalists*? On the nature of capitalists and their tendency to form cartels to monopolize markets, Smith said:

> *We rarely hear, it has been said, of the combinations of masters, though frequently of those of the workman. But whoever imagines, upon this account, that masters rarely combine, is as ignorant of the world as of the subject.*[156]
> (Note: Smith's "masters" are known as "corporate executives," "shareholders" and "owners of capital" today.)

Adam Smith Was Not a Capitalist Zombie. Immediately from his statement above we can see that Smith was quite explicit about his dislike of anybody who disregarded the interests of the general public. Here is Smith on the nature of corporate conspiracies against the public:

> *People of the same trade seldom meet together, even for merriment and diversion, but the conversation ends in a conspiracy against the public, or in some contrivance to raise prices. . . . To widen the market and to narrow the competition, is always the interest of the dealers. . . . The proposal of any new law or regulation of commerce which comes from this order, ought always to be listened to with great precaution, and ought never to be adopted till after having been long and carefully examined, not only with the most scrupulous, but with the most suspicious attention. It comes from an order of men, whose interest is never exactly the same with that of the public, who have generally an interest to deceive and even oppress the public, and who accordingly have, upon many occasions, both deceived and oppressed it.*[157]

Was Adam Smith a Communist? If somebody didn't know who

---

156 Smith, Adam. (1776). The Wealth of Nations. Book I, Chapter VIII.. para 13.
157 Smith, Adam. (1776). The Wealth Of Nations, Book IV, Chapter VIII, p. 145, paras. c29-30.

said that statement above, they might assume it was Karl Marx. That's another example where Adam Smith demonstrated he was not a capitalist zombie. In fact, his books are full of thoughtful, rational, and well-balanced assessments of the virtues *and* vices of capitalism. This is a much more interesting and nuanced perspective of Smith that is rarely presented in college Economics classes today.[158] Here is Smith on the tendency of capitalists to manipulate government policies to enrich themselves at the expense of the general public:

As soon as the land of any country has all become private property, the landlords, like all other men, love to reap where they never sowed, and demand a rent even for its natural produce. . . . Whenever the legislature attempts to regulate the differences between masters and their workmen, its counsellors are always the masters. . . . All for ourselves, and nothing for other people, seems, in every age of the world, to have been the vile maxim of the masters of mankind. . . . Civil government, so far as it is instituted for the security of property, is in reality instituted for the defence of the rich against the poor, or of those who have some property against those who have none at all.[159]

Adam Smith on Taxes to Support General Human Welfare. Many people invoke Smith's name to claim that governments have no right to collect taxes to fund social welfare programs. In making this claim, they ignore Smith's many explicit statements on taxation to fund the collective welfare of a nation, *and not just for homeland defense.* Smith understood that social welfare programs are just as important as roads, infrastructure, and national security because broad-based prosperity and social stability are not achieved through market mechanisms alone. Smith understood that peaceful societies cannot exist without substantial government expenditures to create social stability.

Smith on the need for corporations to pay taxes to preserve the health, welfare, and integrity of their home societies:

A regulation which enables those of the same trade to tax themselves in order to

---

158 This is explored more deeply in my previous book, *Broken Capitalism: This Is How We Fix It.*
159 Smith, Adam. (1776). The Wealth Of Nations, Book V, Chapter I, Part II, 775.

provide for their poor, their sick, their widows and orphans, by giving them a common interest to manage. . . .[160]

Clearly, Adam Smith would not approve of the tax gimmicks and loopholes that U.S. politicians give to their corporate overlords today.

Smith on the principle of progressive taxation on "rents" (*not* income) to assist the poor:

The necessaries of life occasion the great expense of the poor. They find it difficult to get food, and the greater part of their little revenue is spent in getting it. The luxuries and vanities of life occasion the principal expense of the rich, and a magnificent house embellishes and sets off to the best advantage all the other luxuries and vanities which they possess. A tax upon house-rents, therefore, would in general fall heaviest upon the rich; and in this sort of inequality there would not, perhaps, be any thing very unreasonable. It is not very unreasonable that the rich should contribute to the public expense, not only in proportion to their revenue, but something more than in that proportion.[161]

Yes, you read that correctly; the father of capitalism explicitly said that rich people should pay proportionally *more* than poor people to preserve the integrity and stability of their home societies and economies. In fact, for Smith and all classical economists who perceived capitalism through the rational lens of humanism (*not* unsustainable libertarianism or neoliberalism), Smith's comments are obvious common sense.

Smith on how a "whole society" should pay taxes to support public education:

---

160 Smith, Adam. (1776). The Wealth Of Nations, Book I, Chapter X, Part II.
NOTE: This passage is also consistent with the medieval guilds that raised taxes to care for the families of the guild members when they were destitute or otherwise incapable of surviving without community support from the guild. The guilds were highly integrated economic, social, political, and spiritual communities, which had a much more holistic perspective on how to organize communities than Neoliberals do today.
161 Smith, Adam. (1776). The Wealth Of Nations, Book V, Chapter 2, Part II. It's also important to note that he is not talking about income taxation. He explicitly says, "a tax upon house-rents" because he was opposed to any form of income tax because it punishes productive activities.

The expense of the institutions for education and religious instruction, is likewise, no doubt, beneficial to the whole society, and may, therefore, without injustice, be defrayed by the general contribution of the whole society.[162]

In other words, the grotesquely wasteful and inflationary U.S. education system, which is enslaving and choking the life out of entire generations of American college students with mountains of odious debt, should be replaced with a much more cost-efficient public system, just like all the other OECD countries that rank equal or higher than the U.S. in all major education metrics. Then, anybody who is unhappy with the public system can receive vouchers to attend a private school if they want.[163] This would ensure that all Americans receive a solid educational foundation without sentencing them to life-long debtor's prison.

Smith on taxes to support the administration of justice:

The expense of the administration of justice too, may, no doubt, be considered as laid out for the benefit of the whole society. There is no impropriety, therefore, in its being defrayed by the general contribution of the whole society.[164]

Smith on the importance of taxing the people who are closest to where tax revenue will be used:

Those local or provincial expenses of which the benefit is local or provincial (what is laid out, for example, upon the police of a particular town or district) ought to be defrayed by a local or provincial revenue, and ought to be no burden upon the general revenue of the society. It is unjust that the whole society should contribute

---

162 Smith, Adam. (1776). The Wealth Of Nations, Book V, Chapter I, Part IV.
163 Based on analyzing education system performance data across dozens of OECD countries, this approach would be beneficial to all Americans for many economic, cultural, social and geopolitical reasons. And no, the problem is not merely caused by federal meddling in the education system, as many puritanical libertarians claim. Readers interested in learning more about the performance of the U.S. education system and for-profit charter schools compared to the education systems of other countries can read:
Global Governance Scorecard: Eanfar.org
Gini Community Governance System: GiniFoundation.org/kb/community-governance-system/#entity-category-school-university-college
164 Ibid.

towards an expense of which the benefit is confined to a part of the society.[165]

In other words, all federal programs and corresponding taxes and tax loopholes are tyrannical if they only benefit narrowly defined special interest groups, gigantic corporations, or localized communities. If a narrow special interest group, corporation, or local community wants a tax break, then let their local constituencies pay for it and stop stealing wealth from everybody else!

Smith on taxes and public tolls to pay for roads and other infrastructure projects:[166]

The expense of maintaining good roads and communications is, no doubt, beneficial to the whole society, and may, therefore, without any injustice, be defrayed by the general contribution of the whole society. This expense, however, is most immediately and directly beneficial to those who travel or carry goods from one place to another, and to those who consume such goods. The turnpike tolls in England, and the duties called peages in other countries, lay it altogether upon those two different sets of people, and thereby discharge the general revenue of the society from a very considerable burden.[167]

Smith on taxes to pay for public programs and projects that are "beneficial to the whole society":

When the institutions or public works which are beneficial to the whole society, either cannot be maintained altogether, or are not maintained altogether by the contribution of such particular members of the society as are most immediately benefited by them, the deficiency must in most cases be made up by the general contribution of the whole society.[168]

In other words, large projects that have long R&D gestation periods before they can be profitably commercialized, *and that would directly benefit*

---

165 Ibid.
166 Notice that Smith said nothing about privatizing roads with private toll-collecting corporations. We will discuss the issue of privatizing public infrastructure later.
167 Smith, Adam. (1776). The Wealth Of Nations, Book V, Chapter I, Part IV.
168 Ibid.

a large majority of citizens in the country, are justifiable government R&D programs. This includes big infrastructure projects like interstate highway systems and power-generating dams; fundamental scientific R&D into cold fusion and renewable energy technologies; safe Thorium-based nuclear energy; major interstate transportation and utility systems; space travel and related technologies; national defense technologies, among others.

However, taxpayer-funded R&D projects are only justifiable as long as the profits from their commercialization flow back to the taxpayers in a 50-50 partnership with private sector entities selected in a transparent bidding process. This is critical because the corrupt practice of socializing losses and privatizing profits from taxpayer-funded R&D programs is another way that U.S. politicians are stealing wealth from the American people.

Smith on Tariffs. Smith is famous for his "invisible hand" concept, which many people incorrectly assume means that government should never regulate or tax any market activities whatsoever. In reality, Smith was not opposed to tariffs, anti-trust enforcement to prevent market failures, taxes, and government regulations that could be reasonably expected to improve the welfare of a country's citizens and prevent humanitarian crises. For example, Smith on tariffs:

There may be good policy in retaliations of this kind, when there is a probability that they will procure the repeal of the high duties or prohibitions complained of. The recovery of a great foreign market will generally more than compensate the transitory inconveniency of paying dearer during a short time for some sorts of goods. . . . [However] when there is no probability that any such repeal can be procured, it seems a bad method of compensating the injury done to certain classes of our people to do another injury ourselves, not only to those classes, but to almost all the other classes of them.[169]

Our Economic Godfathers Had a Home Bias. I suspect this will surprise many people who have never actually read the complete works of Adam Smith, David Ricardo, John Stuart Mill, or the other

---

169 Smith, Adam. (1776). The Wealth Of Nations, Book IV, Chapter 2, para 39.

Godfathers of Economics, but Smith, Ricardo, Mill and others had a *home bias* perspective on global trade. Why did they have a home bias? Because they understood that *relatively liberal trade* is desirable, but corporations and investors destroy their home economies when they disregard their home societies. This is what many of the largest *American-in-name-only* transnational corporations have done since the 1980s. In fact, this is why Adam Smith invoked his "invisible hand" phrase.

Specifically, Smith assumed capitalists would be loyal to their home country and they would be guided by an *invisible hand*, which he believed would naturally compel them to support their *domestic economies*. In Smith's own words:

> *As every individual, therefore, endeavours as much as he can both to employ his capital in the support of domestic industry. . . . By preferring the support of domestic to that of foreign industry, he intends only his own security; and by directing that industry in such a manner as its produce may be of the greatest value, he intends only his own gain, and he is in this, as in many other cases, led by an invisible hand to promote an end which was no part of his intention....*[170]

David Ricardo, the father of the principle of *comparative advantage*, also said that capitalists should "be satisfied with the low rate of profit in their home country. . . ."[171] Based on their hostility to American workers, tax-dodging, and threats of *corporate inversion*, many of the largest transnational corporations since the 1980s have deviated far from the home bias wisdom that was foundational to the free-trade principles conceived by our Economic Godfathers. (You can learn much more about this topic and the consequences of *transnational cannibalism* in my previous *Broken Capitalism* book.)

Adam Smith Warned Against Harmful Wealth Concentration. Smith assumed that free markets would distribute wealth relatively equitably (i.e., "in nearly the same distribution" and "into equal portions") throughout each national population, which some people conveniently ignore and deny with vague notions of *winners and losers*.

---

170 Smith, Adam. (1776). The Wealth Of Nations, Book IV, Chapter 2, para 9.
171 Ricardo, David. (1821). On the Principles of Political Economy & Taxation. Chapter 7, para 19.

Smith's perception of "winners and losers" did not include wealth gaps that are 5-7 *orders of magnitude* (i.e., up to 10,000,000 times) between the net worth of *average* Americans vs. the wealthiest Americans today.[172] The gap is even larger if we include the *poorest* Americans or the poorest people in all countries, which pushes the wealth gap another order of magnitude higher, i.e., the wealthiest humans have up to 100,000,000 times more wealth than the poorest humans on Earth.

In his 1759 book, *The Theory of Moral Sentiments*, we can see that Smith understood the critical importance of broad-based wealth creation and distribution when he said:

> *The rich . . . are led by an invisible hand to make nearly the same distribution of the necessaries of life, which would have been made, had the earth been divided into equal portions among all its inhabitants, and thus without intending it, without knowing it, advance the interest of the society. . . .*[173]

Smith explicitly warned against squeezing the middle class and allowing wealth to concentrate into the hands of a few. In his own words:

> *No society can surely be flourishing and happy, of which the greater part of the members are poor and miserable. It is but equity, besides, that they who feed, cloath and lodge the whole body of the people, should have such a share of the produce of their own labour as to be themselves tolerably well fed, clothed, and lodged.*[174]

In other words, Smith certainly would never approve of any government that adopts the toxic version of *bank capitalism* that exists in the U.S. and many countries today, which pushes over 80% (and growing) of their populations into debtor's prison *just to maintain a stable quality of life*.

Amorality Check. For people who prefer to keep morality out of

---

172 There is a magnitude range here because net worth can be calculated several different ways.
173 Smith, Adam. (1759). The Theory Of Moral Sentiments, Part IV, Chapter I, pp.184-5, para. 10.
174 Smith, Adam. (1776). The Wealth Of Nations, Book I, Chapter 8, para 35.

economic policy discussions, let's be clear: These are not merely normative or moral concerns. Adam Smith, Smithians in general, and this book discuss some of the problems associated with wealth concentration because excessive wealth concentration *is always* harmful to the integrity and structure of societies, economies, and democracies. There's nothing wrong with becoming *rich*, but when obscene wealth and obscene poverty are allowed to coexist within the same capitalistic economy, it means that economy has deviated far from the original purpose of capitalism, as perceived by Adam Smith.

Should We Blame Rich People?

Somebody recently asked me, "Are you blaming rich people?" Whenever anybody asks me that question, I instantly know they have never read any of my books or detailed articles where I carefully analyze many aspects of our economic and geopolitical systems on Earth today. After one of my friends asked me that question, I wrote the article, "Should We Blame Rich People?" as my response to him.[175] However, that article was too oblique and abstract because I was trying not to hurt his feelings. Since then, as predicted, the socioeconomic and geopolitical problems described in my previous book have become significantly worse. So, let's explore this question in more concrete terms here.

What Does "Rich" Mean? This question is explored more deeply in the *Broken Capitalism* book, but for our purposes here, the phrase "rich" can be defined as "any culturally acceptable multiple of the median net worth in a given society." This definition is important because *human civilization* is shaped and fueled by *human culture*. Human culture is defined and experienced as the *philosophical values* that are *shared by the majority of humans* in each *human society*. A *human economy* is defined and perpetuated by the cultural (philosophical) values that determine which items have *exchange value*. Thus, the puritanical libertarian notion that the denatured principles of an illusory *pure free market* should govern and dominate any human society is a self-destructive fantasy.

What is a *Culturally Acceptable* Level of Wealth? Every society

---

175 See: Eanfar.org/blaming-rich-people

should answer this question based on their own cultural values and *legitimate* economic policies, i.e., policies *approved by a democratic majority of their citizens*. However, a "rich" human with a net worth that is hundreds or even tens of thousands of times higher than the median net worth is probably culturally acceptable in most capitalistic societies today. But in our world today, we have humans with a net worth that is *tens of millions of times* higher than the median *while billions of humans and entire communities disintegrate into poverty*. As the escalating protests in the U.S. and worldwide confirm, these astronomical wealth gaps are not culturally acceptable, nor are they economically or politically sustainable.

Unsustainable Wealth Gaps Create Feudalistic Societies. As the median wealth gap rapidly grows today, puritanical libertarians are dismantling all the social stability programs upon which an increasingly impoverished global population depends. These unsustainable outcomes are only possible within a broken political system that has been captured by a tiny number of super-wealthy creatures. As the labor force participation rate falls, median real income declines, and median debt rises, these conditions are creating societies that look more like medieval feudalism than Adam Smith's conception of market-based capitalism. This is not surprising given that the ideological "precursor" of the *modern* libertarian movement was the infamous plantation slave owner, John C. Calhoun, who fought tirelessly to preserve the lucrative slave trade so that nobody would ever take his "property" (slaves).[176]

---

176 Tabarrok, A., & Cowen, T. (1992). The Public Choice Theory of John C. Calhoun. Journal of Institutional and Theoretical Economics (JITE) / Zeitschrift Für Die Gesamte Staatswissenschaft, 148(4), 655-674. Retrieved from Jstor.org/stable/40751557

NOTE: Nobody will ever know for sure how much James Buchanan was influenced by Calhoun; and whether Buchanan was a racist like Calhoun is irrelevant to the point here. The point here is to indicate that Buchanan's version of "Public Choice Theory" places "property" above all other considerations (just like Calhoun did), even when the gluttonous accumulation of "property" and unabashedly despotic measures to block democratic majority consensus violate deep cultural and social values that are necessary to preserve the institutional integrity and stability of every democratic society and economy. In this particular respect, Buchanan and anybody who subscribes to his theories are deeply misguided and manifestly hostile to consensus-based democracy, which is a recipe for constant social conflict and violent revolution.

Finally, as a former hard-core libertarian myself, I'm still often sympathetic to many libertarian principles when it comes to the role of government and markets in human societies. However, after doing the research for multiple books and over 100 economic and geopolitical articles, it's clear to me that Libertarianism is a woefully inadequate governing philosophy in a world where Continues on next page.

Should We Blame "Rich People" for All These Problems? The problem is not "rich people"; the problem is a political system in the U.S. and U.K. (and several other countries) that allows humans to use their financial wealth to dominate the political system. More specifically, since the late-1970s, there has been a small subset of the wealthy population in these countries that have twisted Classical Liberalism into a rabid form of modern Libertarianism called Neoliberalism, which prioritizes *property* above all other considerations in a society. They equate any form of government intervention (including rational taxation) as *hostile* to their "property," and thus, hostile to their liberty and freedom.

The Parasite Class. That small group of super-wealthy neoliberals has injected massive sums of money into political systems with the explicit intent of hijacking the democratic process. They are explicitly hostile to the fundamental principles of democracy and the *democratic will of the people* because they believe the general population seeks only to steal their property, and thus, must be stopped at all costs. They completely ignore all the systemic problems *they are creating*, which perpetuates the poverty, financial crises, and the conditions that are destroying the middle class in many countries today. Then, they often invoke the phrase "parasite class" to describe the billions of humans who are suffering from their self-serving neoliberal economic policies and destructive corporatocracies.[177]

The Democide Class. As a result of their explicit and persistent hostility to democracy and their corresponding destruction of civil society, I refer to this small subset of super-wealthy, neoliberal humans as the Democide Class. Some people might object to this, but it's common to use the phrase "suicide" metaphorically to communicate the idea that something is self-destructive, but not literally self-murdering. In the same way, "democide" can be used metaphorically to mean anything

---

A.I., structural unemployment, and the convergent interests of gigantic corporations and politicians are creating a feudal global society that makes a mockery of free-market capitalism and all forms of democracy.

177 For detailed evidence and a deeper exploration of the scourges of Neoliberalism and puritanical Libertarianism, read: *Democracy in Chains* by Nancy MacLean, *Dark Money* by Jane Mayer, and many other books in the Gini Book List: GiniFoundation.org/kb/book-list.

that has a tendency to destroy the people and democratic institutions of a country, but not literally murdering the entire population. With their self-serving and shallow economic ideology, the Democide Class is destroying the foundations of democracy by hijacking the institutions of democratic governance and replacing it with autocratic, modern slavery.

The Democide Class Hates Big Government, Except When it's *Their* Big Government. The Democide Class is spawning an ever-expanding police-state government to suppress the rapidly growing mass anger caused by their destructive economic policies. Their police state includes massive military and spying operations to suppress the population in countries they exploit with their natural resource extraction and economically oppressive trade and banking policies.[178] Yet, they have the audacity to claim they are fighting for *human liberty*.

Where Does the Democide Class Get Their "Property"? While the rest of humanity suffers in a slow-burning human apocalypse, the Democide Class myopically equates "liberty" with "property" and essentially shouts *mine-mine-mine* like selfish children fighting over their toys whenever anybody tries to support more equitable and sustainable economic policies. They conveniently ignore the fact that their "property" is usually produced by exploiting the natural resources of countries, destroying the environment, and extracting the financial wealth of large populations with financial engineering scams, which destabilize economies and push a growing global population of economic refugees into deeper debt, poverty and institutionalized slavery every day. Who is the "parasite class" now?

Consistency Check. Anybody can invoke the name of Adam Smith or his *Invisible Hand* to justify their trade and economic policies, but it's reasonable to expect them to be logically consistent. That means acknowledging Smith's explicit guidance to construct societies and economies that are socially and economically sustainable and equitable, which means implementing policies that result in much more evenly distributed income and wealth than we see in the United States and around the world today.

---

178 Use the search feature at GiniFoundation.org to read several case studies that confirm this.

Adam Smith's Perspective on Self-Interest

Adam Smith is often unfairly derided by many people because they take his comments about self-interest out of context. When Smith said, "It is not from the benevolence of the butcher, the brewer, or the baker, that we can expect our dinner, but from their regard to their own interest," he was making a simple, factual observation about the way *every relatively free market works.*[179] A free market cannot function without human free will and personal liberty to pursue the ideas that inspire humans to innovate and prosper through commercial production and exchange.

Self-Interest is Not a Virtue. In that same section of *The Wealth of Nations*, Smith was also making a pragmatic philosophical point about how people should not feel ashamed if they have personal goals and ambition, *but he was not elevating self-interest to the highest virtue of humanity* like Ayn Rand did in her book, *Atlas Shrugged*. Entire generations of capitalists (especially in the financial sector) have grown up believing that Adam Smith really believed self-interest is a moral virtue. They would have been quickly liberated from this self-serving and self-destructive delusion if they had actually read any of Smith's books or attended any of his university lectures on Moral Philosophy, Natural Theology, and Ethics.

It's also useful to see the full context of Smith's "invisible hand" comment because it reveals how *careful and qualified* he was with his words about self-interest.

> *By preferring the support of domestic to that of foreign industry, he intends only his own security; and by directing that industry in such a manner as its produce may be of the greatest value, he intends only his own gain, and he is in this, as in many other cases, led by an invisible hand to promote an end which was no part of his intention. Nor is it always the worse for the society that it was no part of it. By pursuing his own interest, he frequently promotes that of the society more effectually than when he really intends to promote it.*

Self-Interest Alone Does Not Produce Optimal Outcomes.

---

179 Smith, Adam. (1776). The Wealth Of Nations, Book I, Chapter 2, para 2.

Notice in the quote above that Smith clearly uses the words "many" and "frequently" to describe the frequency of positive outcomes *that can* occur from manifestations of human self-interest. He does not use the words "always" or "every time"—he does not even say "a majority of the time." In fact, nowhere in his writings does Smith say that self-interest *always* leads to optimal market-based outcomes. Thus, when puritanical libertarians and neoliberals say that markets driven by human self-interest alone are the only path to an efficient economy, they're distorting the spirit and intent of Smith's entire body of morally-centered, humanistic, socioeconomic scholarship.

Adam Smith Was Not a Heartless Capitalist. Many people have incorrectly assumed that Adam Smith was a heartless capitalist with little or no regard for the plight of the general public. Nothing could be further from the truth. In fact, the first major book that Smith wrote was *The Theory of Moral Sentiments,* in which he stated:

> *How selfish soever [sic] man may be supposed, there are evidently some principles in his nature which interest him in the fortune of others and render their happiness necessary to him though he derives nothing from it except the pleasure of seeing it.*[180]

Adam Smith Placed Ethics Above Economics. When Smith was the Chairman of the University of Glasgow in Scotland, he explicitly prioritized his lecture sequence in the following order: Natural Theology, Ethics, Jurisprudence, *and then* Economics. This sequence was intended to give his students an ethical foundation and societal awareness *before* they learned about Economics.[181] Smith also frequently lectured on the virtues of charity and the need to support people in poverty, orphans, and others who suffer from undeserved misfortune; *and he was not only talking about voluntary charity.* As we have already seen, Smith explicitly endorsed corporate taxation as a necessary fundraising mechanism to take care of destitute humans and stabilize communities, just as the merchant guilds in Europe and China did for hundreds of

---

180 Smith, Adam. (1759). The Theory Of Moral Sentiments, Part I, Chapter I, para. 1.
181 Smith, A., & Wightman, W. P. (1982). The Glasgow Edition of the Works and
Continues on next page.

years before Smith wrote *The Wealth of Nations*.

Adam Smith's Hostility to Government Intervention. Smith's hostility to government intervention was explicitly directed at the harm caused by Mercantilism, which was the dominant economic system on Earth prior to and during much of Smith's life. By definition, Mercantilism is an economic system in which politicians use trade and currency barriers to control all market activity because they perceive international trade as a zero-sum game, rather than a dynamic pie that can be expanded with more economically liberal trade practices. However, there is nothing in Adam's Smith's writings that equates the pragmatic desire for "more liberal trade" with "no government intervention whatsoever!" To the contrary, Smith was not averse to government intervention when it is necessary to correct market failures that reduce sub-national human welfare.[182]

Adam Smith Cared About Socioeconomic Ecosystems. Smith's quotes included in this book are reflective of how thoughtfully he considered the many ways that capitalism impacts human societies. Clearly, he was not merely interested in production and profit; to the contrary, he was deeply concerned about the economic and social ecosystems in which capitalists and corporations operate. He understood that maximizing production and profit are important, *but not sufficient alone*, to maximize human welfare. That's why he believed it is sometimes necessary for governments to intervene in markets to prevent market failures and for corporations to allocate some of their profits (in the form of taxes) to the maintenance of their socioeconomic ecosystems.

The Birth of *Modern* Capitalism. The original field that Adam Smith invented was "Political Economy," which explicitly embraces the *inseparable* relationship between political and economic systems. This systemic duality between political and economic phenomena is woven

---

Correspondence of Adam Smith: 3. Indianapolis: Liberty Press.

182 By definition, a "market failure" occurs in any situation in which the distribution/allocation of some resource is inefficient, often leading to a net loss to unsuspecting stakeholders. Buy/sell walls, bear raids, pump-and-dump schemes, market monopolies, cartels, and many other types of market manipulation (which are only possible when wealth is highly concentrated) are market failures.

throughout all of Adam Smith's work. In fact, this was the perspective that Smith had when he coined the phrase, "capitalism."

The Distortion of Modern Capitalism. Most of the political and sociological analysis that is required to truly understand any socioeconomic system—including capitalism—was pushed to the fringes of the Economics profession by the early 20th Century. This is when gigantic corporations started paying and supporting economists to write *corporation-friendly* textbooks and research reports that focused only on the "economics," while ignoring all the political and sociological dynamics that directly shape real-world economic outcomes. This is like looking at only one side of a coin or one side of a planet and then claiming to know everything about the other side. This one-sided perspective of human existence became the basis for Neoclassical Economics and its neoliberal cousin, which have dominated the Economics profession since the early 20th Century.

What Is the Gini Foundation's Economic Philosophy? Now that we've spent some time verifying Adam Smith's perspective on capitalism in his own words without the propaganda filter of the mainstream media, ideologically captured economists, self-serving politicians and corporations distorting Smith's words, we can see that Smith's perception of Political Economy was based on *humanism*, which places the welfare of human societies above all else. Gini's economic philosophy is aligned with the *real Adam Smith* and the true *humanitarian purpose* of capitalism as he perceived it. We don't really need an *ism* for it, but for those who like labels, we call our particular philosophy Economic Humanism.

In the following chapters we will discuss how Gini's nonpartisan economic philosophy guides the technical development of the Gini cryptocurrency, the corresponding Gini ecosystem, and our public policy recommendations to solve the problems described in Part 1 of this book.

Key Points

- **The Real Adam Smith Was Not a Heartless Capitalist.** In fact, he taught lectures on Ethics, Charity, Theology, Moral

Philosophy and he frequently grappled with the virtues and vices of capitalism like every thoughtful human should.

- **Capitalism Can Be Configured for Humanism or Cronyism.** Capitalism is a socioeconomic operating system for humanity. Like any operating system, capitalism can be *configured* to accomplish a human society's highest priorities; or, it can be configured to serve the interests of a tiny number of corporations and their wealthiest shareholders. If a society's highest priority is to maximize the health, prosperity, and liberty for the largest number of humans, then the puritanical libertarian ideology of pure free markets is not sufficient. Specific public policies must also be in place to ensure the equitable and sustainable distribution of value and wealth throughout an economy.

- **Smithian Economists Take a Holistic Approach.** People following in the tradition of Adam Smith look at all the knowable factors that impact human health and welfare, not merely supply and demand, profit maximization, and national GDP. This holistic approach is the only way to ensure that economic and geopolitical ecosystems are not destroyed by high concentrations of wealth and power.

- Chapter 5 -
Conquering the Tax Monster

"The power to tax is the power to destroy."
— U.S. Supreme Court Chief Justice John Marshall

Why Do We Need to Conquer the Tax Monster? Recall that tax policy is the ultimate battleground where the epic fight for privacy and liberty will inevitably be won or lost in the coming years. The Gini cryptocurrency can give humans power to defend themselves against specific forms of government *and* corporate tyranny. However, technology alone is not enough. *Nonpartisan* citizen power must also be exercised to restore integrity and justice to the tax policies that are oppressing virtually all Americans and much of humanity today. This chapter provides a deeper understanding of the relationship between tax policy, economics, wealth creation and destruction, human poverty and progress, and a nonpartisan solution that would be superior to the existing broken tax system on every measurable level.

To understand how to conquer the Tax Monster, it's important to briefly summarize the relationship between taxes, land values, monetary inflation and their socioeconomic impact on human societies.

Nature's Gift to Humanity. When people talk about the value of "real estate," they're really talking about two concepts: the value of land and the value of the *property* (buildings, structures, facilities, etc.) *on top* of the land. Land and property are two separate components when we calculate the entire value of real estate. That means we can easily separate land and property and treat them differently for tax purposes. Land is the Earth, which is Nature's gift to humanity. Property is everything else that humans build on top of land.

What is the Value of the Earth? Land is extremely valuable—more valuable than many people realize. Only 6% of the 2 billion acres of U.S. land is developed, which means land is the most untapped natural resource in the country.[183] One tiny island alone—Manhattan Island in New York City—is only 23 square miles, but it's worth about $1.74 *trillion* today.[184] That's nearly 50% of the entire U.S. Government's annual budget.

Squatters Create Artificial Price Inflation. Land speculators are people who buy undeveloped land and squat on it for years until the price rises, then they sell it for a nearly risk-free profit. They don't add any buildings or facilities (property) to the land; they just squat on it. Thus, they add no value to the land. Squatters hurt communities in many ways. In particular, squatting on land causes urban blight and artificially restricts the supply of land, which artificially increases home, apartment, farmland, and commercial office rental prices. Artificially higher rental prices create price inflation, which increases the cost of everything in an economy.

Squatters Suck Wealth from Everybody in a Country. The inflation created by squatters sucks wealth from everybody in a community, including all the families and businesses living there. This is because every business must charge the families more for everything (food, entertainment, utilities, cars, clothes, etc.) just to pay the artificially high rental prices to the squatters. Now, consider that every community, city and state throughout a country suffers from the same squatter and price-inflation problem. This is a massive multi-trillion-dollar problem. In fact, this problem is the fundamental source of nearly all the so-called *natural* inflation throughout every economy; but in reality, there's nothing *natural* about it.

Squatters Add Zero Value to an Economy. This may seem sacrilegious to some people who have not taken the time to analyze how and why land ownership is separate from property ownership and how these two components of real estate impact an economy differently. To prove that squatters add zero value to an economy, it's useful to visualize how real estate is comprised of two completely separate

---

183 Larson, W. (2015). New Estimates of Value of Land of the United States. US Bureau of Economic Analysis, 30.
184 Barr, J., Smith, F. H., & Kulkarni, S. J. (2018). What's Manhattan worth? A land values index from 1950 to 2014. Regional Science and Urban Economics, 70, 1–19. Doi.org/10.1016/j.regsciurbeco.2018.02.003

ingredients: land and property.

Recall that property (buildings, factories, etc.) is built *on top* of land, but property is separate from land. Now, consider *who actually creates the value* embodied within every parcel of land. . . .

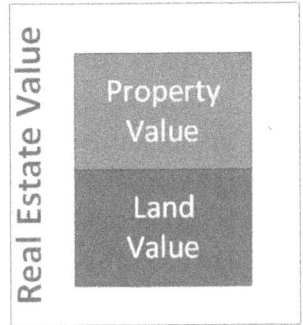

Community/Taxpayers Create Land Value by:

- Building infrastructure & utilities (gas, electric, water) on and around the land.
- Building schools, parks, facilities, and other common areas on and around the land.
- Enforcing the rule of law on and around the land.
- Providing national defense on and around the land.
- Entrepreneurs & capitalists build valuable business ventures, create jobs, innovate, etc., which attracts people to live on and around the land.
- The community collectively generates economic growth, prosperity, and cultural richness, which attracts more people to live on and around the land.
- Families invite other families to live on and around the land.

Squatters Create Land Value by:

- No meaningful value whatsoever—buying and squatting on a parcel of land does absolutely nothing to create or increase the value of land. To the contrary, squatters suck up the value created by the community/taxpayers, but they add no value to the land or community whatsoever. And their squatting artificially reduces the supply of land, which artificially inflates the price of land and everything else in the economy. (Remember: *Property developers* are very different from land squatters. Property developers actually *develop* the land.)

Land Is the *Natural* Foundation of Every Economy. If you're still not convinced of the harm caused by squatters, the image below illustrates the fundamental structure of every economy on Earth. Everything in human civilization must be built on top of land. That means squatters have leverage over all of human civilization, *but they did not create the land or any value on the land.* In fact, land is the only element of an economy that is fixed, scarce, *and cannot be replenished by humans.*[185] Land is the source of all material wealth, dietary nutrition, and human health, but when squatters hoard this natural resource, the price inflation that they create bubbles up throughout the entire pyramid of the economy, unjustly sucking wealth from every human within the economy.

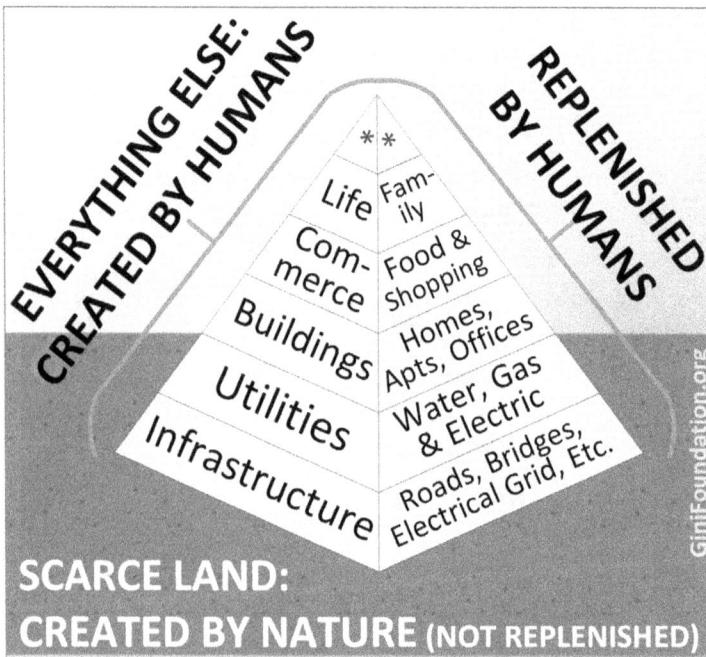

One Simple Tax to Replace Them All

Today, Americans are drowning in taxes: income taxes, sales taxes, capital gains taxes, estate taxes, excise taxes, payroll taxes, social security

---

185 Artificial, manmade islands are not large enough to significantly impact the value of natural land.

taxes, property taxes, gasoline taxes. . . . Given the USG's $22 trillion debt and $1 trillion annual budget deficit, the taxes are going to get much worse. Currently, all those taxes amount to only about $3.6 trillion to the government, which is only two times the $1.74 trillion value of Manhattan Island. And recall that approximately $1 trillion is wasted due to the bureaucratic cost of collecting the income taxes alone. Wouldn't it be great if we could replace all those confusing, time-consuming, and absurdly wasteful taxes with a single, simple, sustainable tax?

Conquering the Tax Monster—Part 1: The Land Value Tax. By separating Nature's gift to humanity (land) from humanity's productive output (property), we have a simple way to solve this multi-trillion-dollar problem: a Land Value Tax (LVT).[186] This is not the same as a property tax. In fact, an LVT is much simpler than a property tax, but the LVT economically incentivizes people to invest in land *improvements now* rather than speculating and squatting on land to sell it years later. Given that Manhattan Island alone is worth $1.74 trillion, there's a lot of value locked up in that land, but only a tiny number of people actually benefit from that value today. The LVT resolves that problem.

Conquering the Tax Monster—Part 2: The Public Land Trust. Instead of hoarding a natural resource that they did not create, people who need unused land for any purpose should be required to lease it from a Public Land Trust (PLT), which can and should be administered by fully automated blockchain-based smart-contracts. That would eliminate the corruption often associated with politicians who use their control over national resources to reward their cronies and patrons. Additionally, a "public trust" is the most appropriate legal structure because land is fundamentally a public natural resource—a gift from Nature. Thus, nobody should be able to control and exclude others from using any land *unless* they're actually leasing it from the PLT for a productive purpose that adds meaningful and sustainable economic value to society.

Why Does the LVT Incentivize Us to Improve the Land? In an LVT-based tax system, the amount of LVT paid by each land *leaseholder*

---

186 The LVT has been known for centuries and it was supported by Adam Smith and most enthusiastically by Henry George in the late 19th Century. For those interested in the deeper Continues on next page.

is the same for all leaseholders within the same neighborhood, regardless of whether they are using the land for a productive purpose or not. In this case, "productive purpose" means the leaseholder must use the land to generate revenue from some kind of for-profit *or* nonprofit activity. Thus, the land lease is an operating expense, which must be offset by revenue generated by delivering some kind of valuable product or service to society. With the LVT, it doesn't make economic sense for a leaseholder to speculate and squat on land without improving it and using it productively because the land value is taxed regardless of whether they're generating revenue from the land or not. This eliminates the artificial price inflation, which eliminates the economic incentive to squat on the land.

Break the Psychological Addiction to Land Ownership. Some people have a hard time understanding the simple LVT principle because their brains are so addicted to the concept of "land ownership." The human brain can be addicted to many things, including heroin, but that doesn't make it good for us or society. It's important to not get sucked into the fallacy that "this is how we've always done it; so, people are not going to change." That's an irrational, lazy excuse. The current Tax System 2.0 that we've had since 1913 is decrepit and sinking our country. It's time to upgrade to the LVT-based 3.0 version.

Visualizing How the Land Value Tax Works. The image below illustrates how the simple LVT works. Each square represents a separate parcel of land. The squares on the left have no buildings (property) on them; so, they're "undeveloped land." The squares on the right have buildings (property) on them; so, they're "developed land."

economic principles of the LVT, I highly recommend Henry George's book, *Progress & Poverty*.

Undeveloped Land	Developed Land
Land Value: $100,000 Property Value: $0 Annual Tax: $10,000	Land Value: $100,000 Property Value: $50,000,000 Annual Tax: $10,000
Undeveloped Land	Developed Land
Land Value: $100,000 Property Value: $0 Annual Tax: $10,000	Land Value: $100,000 Property Value: $200,000 Annual Tax: $10,000

Land Value Tax Visually Explained   GiniFoundation.org

The LVT Creates Healthy Incentives to Build Sustainable Communities. Notice all the squares have the same *land* value ($100,000), but the developed land properties have different *property* values. These property values represent the *improvements* that the land *leaseholder* (*not* landowner) has made to the land. Improvements increase the value of the land and many other assets in a community. For example, building houses and supermarkets attracts new residents and other businesses to the community. Then, as more people are attracted to the community, this increases demand for *developed land*, which *naturally* increases the price of all assets through *the natural* supply and demand of *productive assets. This is real wealth creation*, not artificial scarcity caused by unproductive feral land hoarded by wealth-sucking squatters.

The LVT Does Not Tax Property Appreciation. It's important to remember that in an LVT system the increasing value (appreciation) of land *improvements* (property) *is not taxed* because that would create an economic *disincentive* to improve the land, which is not economically rational. Additionally, the undeveloped land leaseholder pays the same tax as the developed land leaseholder because the land value is fixed within the same neighborhood. Again, this incentivizes the undeveloped land leaseholder to improve the land quickly or stop squatting on it and let somebody else lease it who wants to improve the land and generate ongoing income from it.

Squatters Destroy Economies & Societies

The Tax System Today Produces a Double-Penalty. It's nearly impossible to overstate how tragically unjust, backward, and wasteful the current income tax system is. For example, the existing tax system takes economic land value (recall that *all* land value is created by entire communities) and transfers that economic value to individual squatters who produce literally none of that economic value at all. *Then,* the current system taxes labor and capital when they engage in *productive* activity! This is completely backwards and penalizes capital and labor for producing wealth; while rewarding squatters for producing absolutely nothing.

Squatters Create Economic Waste & Unemployment. Because financial capital is much more abundant in an LVT-based system, the unemployment rate with the LVT would be much lower because the entire economy is no longer suffering from multiple forms of inefficiency. For example:

- Inefficient use of space and urban sprawl caused *directly* by unproductive squatter land in the middle of urban areas.
- Economic inefficiency caused by wealth flowing out of the general economy into the pockets of unproductive squatters.
- Economic inefficiency of financial capital being locked up in unproductive squatter land.
- Unemployment caused by inefficient capital deployment to speculative squatter land instead of factories and other productive assets and activities.
- The entire chain of national and global economic inefficiencies

caused by the reduction in productivity throughout an entire global economy due to all the factors above.

Boom & Bust Cycles Are *Not* Natural. Contrary to economic *orthodoxy*, there is nothing *natural* about economic bubbles, painful recessions and depressions. It is not *natural* when these events lash our economic system and shift wealth from the middle class to the rich through bank bailouts and corporate welfare every time the bubbles pop. These events are *directly* caused by speculative malinvestment in land. In fact, bubbles occur because of the following economic distortions.

Squatters Add a Speculative Premium to Every Transaction. The price of every land transaction is inflated with a speculative premium based on what the squatter *hopes* the land will be worth at some point in the future. This premium is a totally arbitrary amount, based primarily on the emotions of the transacting parties and the arbitrary calculations of land appraisers, which often collude with sellers to inflate the premium so they can extract more profit from each transaction. This is in stark contrast to the way all other productive assets are evaluated. For example, the sale price of a company is based on the *productive output* of its assets, net income, ongoing cash flow, and many other factors that are directly linked to creating real-world economic value. In fact, the price of all non-speculative products and services is derived from the *utility value* they provide, not an arbitrary premium and prediction about a future sale.

Squatters Inflate the Money Supply & Debase the Currency. Since virtually all real estate is purchased with credit, the speculative premium gets embedded into every real estate loan, which inflates the size of the loan. Because the banking system is a fractional reserve system, banks create money out of thin air every time they issue a loan. This inflates the money supply, which artificially increases the price of real estate and everything else throughout an economy, thereby fueling the bubble and debasing the country's currency.

Squatters Transform the Economy into a Ponzi Pyramid. Nearly all real estate loans are collateralized by . . . real estate! Thus, speculative land squatting continuously and artificially expands the base of collateral throughout the entire economy. This incentivizes banks to loan more money to real estate speculators instead of innovative entrepreneurs and factory operators. For the squatter, this is wonderful because they can create a pyramid of debt that enables them to keep gobbling up more and more land, then watch the land price artificially inflate, then siphon off the difference between the purchase and sale prices without doing any work at all to create that increased land value. As long as the Ponzi scheme keeps pushing land prices up, the game continues. In the meantime, this pyramid of debt spreads throughout the entire economy, inflating the money supply and creating bubbles.

Squatters Infect Other Industries with Bubbles. Because land is literally the foundation of everything in human civilization, real estate bubbles caused by artificially inflated land prices quickly spread to other industries. This occurs because real estate bubbles increase the price of rents throughout the economy. It also artificially increases demand for goods and services used in the real estate industry. When the bubble pops, it results in massive job losses in all those related industries.

Some people say, "Well, at least the bubble created jobs for a while." That's just an illusion: Without all the economic inefficiencies caused by squatters, the economy would operate at nearly full employment without any bubbles at all because financial capital and land would be available to create many more jobs *sustainably.*

Squatters Hurt the Poor & Middle Class. Because squatters debase the currency by inflating the money supply with a pyramid of fractional reserve debt, they hurt the poor and middle class the most because those people usually have no way to protect their wealth from inflation. The poor and middle class depend on wages to survive, but the value of their wages is continuously eroded by the squatter's currency debasement.

In the meantime, wealthy squatters and sophisticated investors can afford to pay investment advisors to diversify their portfolios and hedge their assets against inflation, but the poor and middle class have no way to do this. High-quality hedge fund managers can't offer their services to the poor and middle class due to financial regulations and because it's not profitable to manage small portfolios. So, the wealth of the poor and middle class evaporates while speculators and other wealthy people are able to gobble up all the assets throughout the economy, which slide ever-farther out of the reach of the poor and middle class.

Squatters Are Directly Responsible for Economic Collapses. All bubbles inevitably pop, which creates economic recessions and depressions. When this happens, the value of the land crashes back down to its *real economic value to society*, and then the vicious cycle and speculative bubble starts over again. Wall Street is merely a facilitator by providing the liquidity (credit) and exotic derivatives to slice and dice the Ponzi pyramid of debt created by the squatters. Then, when the economy crashes, governments use taxpayer funds to bailout the banks that fueled the speculators' Ponzi debt. By then, the smart squatters have sold the land they were squatting on before the bubble pops and the banks get free money from taxpayers.

Who Loses from this Ponzi Scheme? The general public suffers the most from devastated communities, debased currencies, obliterated retirement portfolios, and higher taxes used to prop up a Ponzi scheme that destroys their quality of life and wealth.

Squatters Gradually Disenfranchise the Poor & Middle Class. By gradually eroding the value of poor and middle-class wages and destroying their wealth with an endless cycle of booms and busts, squatters reduce the financial power of the poor and middle class. Numerous studies have been conducted that reveal how money dominates the U.S. political system; thus, the diminishing financial power of the poor and middle class accelerates their loss of democratic influence over the laws and regulations that control their lives.[187,188,189,190] People with wealth today may not care about this, but wealth is often lost or depleted over time. So, our children and their children may not have the same level of wealth. Should we tolerate squatters disenfranchising our grandchildren, too?

---

187 Taibbi, M. (2014). The Divide: American Injustice in the Age of the Wealth Gap. NY: Spiegel & Grau.

188 Mayer, J. (2017). Dark Money: The Hidden History of the Billionaires Behind the Rise of the Radical Right. Anchor.

189 Kaiser, R. G. (2010). So Damn Much Money: The Triumph of Lobbying and the Corrosion of American Government. New York: Vintage.

190 Hacker, J. S., & Pierson, P. (2011). Winner-Take-All Politics: How Washington Made the Rich Richer—and Turned Its Back on the Middle Class (unknown edition). NY: Simon & Schuster.

Fractional Reserve Banking Is Sustainable *Only* When Loans Are Used for Productive Assets. This is because productive assets are the basis of *real* wealth creation. As long as the total rate of *real* wealth growth (factories, new technologies, real production, etc.) is equal to or higher than the rate of money supply growth, then inflation does not erode the value of a currency.[191] This is because the numerical unit value of aggregate wealth (measured in currency units) expands to absorb the numerical unit growth of the currency supply. As loans are repaid, they *self-extinguish* the money they previously created, which reduces the money supply and results in a relatively steady-state economy.

Squatters Cause Ponzi Money & Phantom Wealth to Clog Up the Banking System. In contrast to a steady-state economy, speculative land transactions that don't create any productive output do not create wealth; they simply spawn endless speculative premiums, debt pyramids, and monetary inflation that are falsely *recorded* as wealth on a squatter's balance sheet. Then those debts and corresponding Ponzi money clog up the banking system for decades (or until the loans are eventually repaid), which means it takes many years to self-extinguish the underlying money supply that was pumped into the system when the loans were created. All that Ponzi money is merely phantom wealth that rapidly deflates or violently evaporates as soon as the bubble pops.

The LVT Creates Market Stability

The LVT Eliminates Real Estate Bubbles. From a technical Economics perspective, purchasing something (e.g., real estate) merely because we want to sell it at a higher price in the future *is not a commercial transaction*; it's a speculative *gambling transaction*. That means it's not a sustainable economic activity because market bubbles are, by definition, caused by an over-accumulation of speculative transactions within an economy. However, with the LVT, real estate markets grow more sustainably and there are never any real estate bubbles because the real estate speculation that fundamentally causes toxic real estate bubbles is

---

191 I call this phenomenon the "Law of Money-Value Creation," which we will revisit in a later chapter when we discuss Gini's monetary policy.

economically impossible in an LVT-based economy. Additionally, because real estate is the largest driver of economic activity throughout an economy, all other industries that are based on the real estate market grow more sustainably, too, in an LVT-based economy.

What About Stock & Bond Market Bubbles? Speculation in any market can drive up the price of assets in that market. However, there are profound differences between speculating in man-made assets like stocks and bonds vs. speculating in land and all other natural resources:

- **The total quantity of land on Earth can never be increased**; so, it's price cannot be moderated by increasing its supply to absorb increased demand. Thus, whenever squatters hoard the supply of land, that artificially increases the price of all land, land-based collateral, and all products and services that depend on land—virtually everything in human existence. That artificially expands debt throughout an economy, leading to all the problems explained previously.

- **The price of stocks, bonds and other man-made assets is based on *real, ongoing cash flows*.** For example, a company's stock price is dominated by its P/E Ratio, which is anchored to the company's actual net income. The price of a bond is anchored to the issuer's ability to repay the debt principal and interest, which is based on the issuer's income or ability to collect taxes from citizens. Nobody would buy a stock or bond if they didn't have confidence in the underlying cash flows of those instruments, which reflect real-world economic activity and wealth flows.[192]

- **Speculating in stocks and bonds can't deprive humans of existential necessities** like housing, food, and water because stock and bond speculation doesn't generally cause the price of existential necessities to skyrocket. Of course, speculating in *specific* stocks and bonds issued by real estate and natural

---

192 The U.S. Treasury Bond market is a rare exception to this rule because it is currently propped up by the U.S. Dollar's global reserve and petro-currency status, but that exception will not last forever. Eventually, the USD will collapse and bring down all USD-denominated assets with it, including U.S. Treasury bonds.

resource extraction companies within today's broken economic system that allows gigantic corporations to monopolize and artificially restrict the supply of natural resources *can* impact the cost of capital for their operations. That *can* impact the price of the natural resources they extract and distribute in global supply chains. But that would be impossible in an LVT-based system that prevents the monopolization of *any* finite natural resource, which we will discuss in more detail later.

LVT-Based Land Appreciation Benefits the Entire Community. It's important to understand that land *does appreciate* in an LVT-based system, but that appreciation is based on *natural* supply and demand of *productive assets*. As people and businesses are attracted to an LVT-based community, the prices of productive assets (including *productive land*) within that community will *naturally* rise in response to the law of supply and demand. This is *real asset appreciation*, not artificial squatter price inflation. This is a profound distinction because *real* asset appreciation benefits the entire community in multiple ways. Specifically, in an LVT-based economy:

(1) The supply and demand of *productive* assets is optimized because there's no artificial scarcity, *dead-weight loss*, or price inflation.
(2) Asset market prices reflect their true utility value to society.
(3) Asset appreciation represents *genuine wealth creation*, which increases the total stock of capital throughout the community.
(4) Capital is spread more evenly throughout the community, which creates a more sustainable economic ecosystem.
(5) The appreciated land is taxed at the same rate as it was before the appreciation, but because the community *worked together* to make land more valuable through their collective actions to improve the community and attract a larger population (thereby increasing the demand for land), the increased land appreciation and corresponding tax revenues that go to the Public Land Trust are *distributed back into the community*. This creates a virtuous cycle of community improvements to land and property, which increases the value of land and property, which increases the tax revenue collected

(*on land only*), which is re-invested back into the community in the form of better education systems, healthcare systems, infrastructure, renewable energy projects, cultural programs, and whatever the community democratically decides is best for the community.

(6) Economic stability, more broadly shared prosperity, dramatically lower market volatility, and more sustainable communities reduces ethnic tensions, which creates more peaceful communities. In other words, in an LVT-based economy, land appreciation benefits the entire community, not only a few rich squatters.

"The LVT is Great! Why Isn't It Already Implemented?" This is a common question. Remember the image of the economic pyramid sitting on top of the land presented earlier? In the dysfunctional, often corrupt, and completely unsustainable economy that we have today, landowners are the most powerful special interest group because they control the one natural resource that gives them leverage over everything in human civilization. So, many of them have no interest in giving up their systemically destructive wealth and power, which is *illegitimate and unsustainable* wealth and power if it's based on monopolizing public natural resources. Because of this, they conveniently ignore many important facts and realities, including:

- *All* land is produced by Nature, not squatters.
- *All* land was originally stolen from native humans. Wealth-sucking squatters amplify that original sin by destroying human communities and economies.
- *All* land value appreciation is created by taxpayer-funded community development, which increases the value of the land and property around the land; thus, squatters receive *unearned* capital gains from the appreciation of an asset that they didn't create.
- The price squatters pay for land is almost always far less than the land is worth in a productive LVT-based economy.
- *All* squatter land is a net loss to society.
- In an LVT-based economy, homeowners and building owners still own their property *on top* of the land. Thus, *the LVT does not*

steal anybody's wealth.

- The LVT could be implemented gradually over several years to ensure a smooth transition. This would be a period of phasing out *all existing taxes* and phasing in the single LVT in a tax-neutral way until the transition is complete.

- Homeowners and building owners who already use their land productively would actually see *a reduction in their total tax burden* because the LVT would be *much less* than they're currently paying for all their cumulative existing taxes.

With the LVT, Does the Government Confiscate Land from Pre-LVT Landowners? After a society switches to an LVT-based system, the government doesn't confiscate the land; it simply imposes a democratically determined rent on all land. If a pre-LVT landowner is using land productively and generating income from the use of that land (e.g., farming, building skyscrapers, shopping centers, etc.), then the LVT land rent would be *functionally* no different than a periodic property tax today. However, if the pre-LVT landowner is not using the land productively, then the LVT land rent is still required. This makes it unprofitable to squat on unproductive land. Thus, the LVT *leaseholder* is incentivized to either develop the land so that it's generating income that can be used to pay the LVT, or get off the land and allow somebody else to use it productively.

What About Homeowners? Homeowners today pay property taxes already, in addition to many other taxes. Their property taxes would simply be converted to the single LVT, all their other taxes would be eliminated, and they would stay in their homes without any inconvenience whatsoever. Their total tax burden would be much lower. Homeowners with equity in their homes would still keep the equity, but it would be equity *in the property* on top of the land, not the land itself. So, if they sell their homes after the LVT is implemented, they would literally be selling their "homes" *only*, not the land beneath their homes. Land—Nature's gift to humanity—should never been "owned" by any human or corporation in the first place!

Where Will Rich People Invest Their Money If They Can't Invest in Land? Real estate today is comprised of land *and* property.

With the LVT, real estate would simply be property *on top* of the land, but not the land. Real estate *property* would still be bought and sold just like it is today. The more interesting question is: "Where else will the largest land squatters find a nearly risk-free asset in which to park their USD billions?" Aha! This question reveals one of the greatest virtues of the LVT: They will have to invest in the *real economy*: factories, innovative startup companies, quantum computers, clean energy technologies, cures for cancer, cures for aging and death—*anything but land.*

The LVT Optimizes the Output & Efficiency of Capitalism. In the broken economic system that we have today, *parking* financial wealth in unproductive land removes that wealth from circulation and deprives the economy of the propulsive force that wealth would otherwise provide to human civilization. In contrast, the LVT naturally incentivizes rich people to put their wealth to productive use, which optimizes every socioeconomic value-creation process within every capitalistic society.

The LVT Is a Single Tax to Replace All Other Taxes. Because land is so ubiquitous and all of human civilization depends on land, it does not require a high land tax to generate huge tax receipts for the government. Thus, a single, small land tax would easily replace all other taxes throughout the economy. This includes eliminating income taxes, payroll taxes, capital gains taxes, sales taxes, excise taxes, gas taxes, property taxes, and all other taxes.

The LVT Eliminates the IRS

The LVT would require only a relatively small federal LVT Department to collect all LVT taxes. What does a "relatively small LVT Department" actually mean in real-world terms?

- The LVT would not require a worldwide police state apparatus to hunt down tax dodgers all over the planet because land can't be moved or shifted to tax havens or hidden with cryptocurrencies; i.e., it's easy to enforce the LVT. Thus, the inhumane and evil "diaspora tax" would be eliminated.
- The "$450 billion in unreported taxes" reported in the Mercatus

study would drop to "$0 in unreported taxes" because it's impossible to evade LVT taxes.

- The "$1 trillion in tax compliance costs" would drop to nearly $0 because the LVT makes tax evasion impossible and "compliance" could be completely automated. In fact, the LVT would be a predictable, recurring quarterly bill, which could also be fully automated.

- Politicians would have no legitimate excuse to block citizens from using cryptocurrencies. (More on this later.)

- Politicians and gigantic corporations would no longer be able to fleece the American public with endless tax scams and loopholes.

- Corrupt politicians and federal officials would no longer be able to use the IRS as a political weapon against their opponents, civil rights activists, or as a system of tyrannical mass social control.

- Americans would be able to do business in many countries again without being treated like Ebola carriers. In contrast, today most countries either completely reject Americans or *quarantine* them in ring-fenced regulatory boxes so that the IRS doesn't harass their governments whenever the IRS is hunting an ever-expanding population of *tax criminals*.

- Armed IRS agents would never be able to kick down a citizen's door and point guns at them for not disclosing their most intimate personal secrets during tax season; and "tax season" would be abolished.

- The entire 74,000-page Tax Code monstrosity could be burned in a 25-foot-tall bonfire. Americans could call that day "Freedom Day," which could be celebrated as a national holiday every year thereafter. For most Americans, this day would be just as emotionally liberating as it was for Germans on the day the evil Berlin Wall collapsed.

- The phrase "1040 tax return" could be purged from the English dictionary.

- The concepts of "corporate and personal tax returns" would no longer exist because all taxes would be replaced with a single, fair, equitable, simple, and sustainable LVT.

- For the first time in over 100 years, the American people would begin to experience real economic freedom.
- The success of the LVT in the U.S. would politically compel politicians in all countries to follow the U.S.' lead, which would liberate the rest of humanity from their own squatter and tax tyranny.
- The U.S. Bureau of Land Management (BLM) already performs a large portion of the activities required for the LVT.[193] Thus, either a small Public Land Trust Agency (PLTA) could be established or the existing BLM could be upgraded to perform all the LVT tax processing activities by increasing the BLM budget by only $100–300 million.[194] That means the entire $11.5 billion IRS budget could be eliminated and all federal LVT activities could be managed with a budget that is 10 times smaller than the IRS' budget today.

What Is a *Good* Tax?

What Characteristics Should an Economically Efficient Tax System Have? Adam Smith discussed the issue of taxation extensively in *The Wealth of Nations*. In fact, he was highly specific about how to evaluate whether a tax is economically efficient or not. To summarize Smith's four "Cannons of Taxation," an effective tax system has the following characteristics:[195]

- Minimal impact on producers & consumers
- Cheap & easy to collect
- Predictable revenue for government & predictable and

---

193 According to the BLM website, "The BLM manages one in every 10 acres of land in the United States, and approximately 30% of the Nation's minerals. These lands and minerals are found in every state in the country and encompass forests, mountains, rangelands, arctic tundra, and deserts."

194 For perspective, $300 million is the size of the Federal Trade Commission's annual budget, which pays for the FTC's approximately 1,000 full-time equivalent positions (employees).

195 Smith, Adam. (1776). The Wealth of Nations. Book V, Chapter 2, paras 24-28.

understandable for citizens

- Fair & equitable for all members of society

Based on those four principles, let's evaluate all the most common types of taxes using an efficient scorecard. After the scorecard below there's a succinct summary of how each score is computed and why the LVT scores so much higher than all other forms of taxation.[196]

Adam Smith's Cannons of Taxation Scorecard

	Low-Impact	Cheap & Easy	Pre-dictable	Fair	Total Score
Income Tax (Progressive)	2	0	1	3	6
Income Tax (Flat)	1	4	1	1	7
Capital Gains Tax	4	1	1	5	11
Sales Tax	1	1	1	0	3
Import Tariff	0	4	0	0	4
Property Tax (Land & Buildings)	3	3	4	2	12
Land Value Tax	**5**	**5**	**5**	**5**	**20**

Source: GiniFoundation.org

Income Tax (Progressive): Penalizes productive activity. Terribly tedious and inefficient to calculate and collect. Government revenue is not predictable because taxpayers' income can fluctuate wildly year-to-year. Progressive taxes are relatively fair according to the "ability to pay" principle, but some people don't accept that and believe everybody should pay a flat tax.

Income Tax (Flat): Penalizes productive activity. Much easier to calculate and collect compared to progressive taxes. Government revenue is not predictable because taxpayers' income can fluctuate wildly year-to-year. High negative impact on poor and middle class (the vast majority of every population) because it adds to the cost of basic

---

196 This scorecard approach was inspired by an excellent lecture, "Understanding Economics," presented by Lindy Davies at the Henry George School of Economics. I highly recommend visiting their website at: Hgsss.org.

necessities, which is regressive on low and fixed incomes. Thus, it's not fair from the "ability to pay" perspective because it creates a greater hardship on the poor and middle class.

Capital Gains Tax: Does not penalize production because capital gains come from unearned income. Tedious to calculate and collect because status of income and assets is often complex and politicized. Government revenue is not predictable because asset values and returns can fluctuate wildly. Very fair from the "ability to pay" perspective because it taxes unearned income and has a low impact on the poor and middle class.

Sales Tax: Penalizes consumption, which reduces total economic activity. Can be difficult to collect due to various merchant evasion tactics and collusion between merchants and customers. Government revenue fluctuates wildly as every merchant's sales fluctuate due to seasonal and competitive dynamics. High negative impact on poor and middle class (the vast majority of every population) because it adds to the cost of basic necessities, which is regressive on low and fixed incomes. Thus, it's not fair from the "ability to pay" perspective because it creates a greater hardship on the poor and middle class.

Import Tariff: High negative impact on production and consumption because it adds to the cost of both. Relatively easy to collect because ports and borders are easier to monitor than individual merchants and taxpayers. Government revenue is not predictable because imports are highly variable for many reasons. High negative impact on poor and middle class because it adds to the cost of basic necessities, which is regressive on low and fixed incomes. Thus, it's not fair from the "ability to pay" perspective.

Property Tax (Land & Buildings): Relatively low impact on production and consumption because the tax can be allocated across a broad range of income-producing goods and services. Relatively cheap and easy to collect because land and buildings can't be moved to evade the tax. Government revenue is much more predictable than other existing taxes. Not very fair because it penalizes productive activity associated with property development, erecting buildings, creating shopping centers, etc., all of which are productive activities that add value to communities.

Land Value Tax (LVT): Zero impact on production and consumption because *land rents* don't affect the price or supply and demand of any merchant's goods or services.[197] The LVT is the easiest and cheapest tax to collect because land is easy to find and assess. In contrast to wildly fluctuating sales and income taxes, government revenue from the LVT is highly predictable because land values are reassessed every x years, which means all LVT payments can be fully automated. Since the LVT is based on a public natural resource and is for the benefit of all citizens, it increases the living standards of the poor and middle class without impacting the wealthy (except for squatters) because the LVT replaces all other taxes, including existing property taxes, which the wealthy already pay. So, the LVT is by far the fairest, most efficient and most equitable form of taxation.

Adam Smith on the LVT. It's clear that Smith understood the mechanics of how land prices are entirely driven by community development, not land squatters; and that land taxes should constitute a majority of all government tax receipts. Smith's 18th Century language might be a bit challenging for modern English speakers to quickly decipher; so, I've inserted some commentary in brackets to make it easier to understand.

Ground-rents are a still more proper subject of taxation than the rent of houses. . . . [Land taxes are more economically efficient than property taxes.] The ordinary rent of land is, in many cases, owing partly at least to the attention and good management of the landlord. A very heavy tax might discourage too, much this attention and good management. [At least property

---

197 This is a well-known virtue of the LVT based on the technical phenomenon of price-elasticity in competitive markets. Specifically, this phenomenon is caused by two factors: (1) In any neighborhood, all merchants pay the same LVT; so, they can't move to another location within the same market to reduce their LVT; and (2) in every competitive market, if a particular cost of a good can be eliminated to gain market share, then it inevitably will be eliminated by one or more merchants to gain market share. That drives out all excess costs from every competitive market. These factors mean the LVT can never be profitably passed on to consumers because there will always be merchants willing to absorb the LVT to offer the lowest price to gain market share. Thus, any merchant that tries to pass on the LVT will lose market share and ultimately go bankrupt. We can see that Adam Smith understood this when he said, "A tax upon ground-rents would not raise the rents of houses." Smith, Adam. (1776). The Wealth of Nations. Book V, Chapter 2, para 74.

developers and landlords earn their money by providing a meaningful service to the residents living in their properties. So, don't tax their properties too heavily.]

Ground-rents [land rents], so far as they exceed the ordinary rent of land [when land prices rise above the utility value of farm land], are altogether owing to the good government of the sovereign [taxpayers and good government make the value and price of land rise], which, by protecting the industry either of the whole people [government provides infrastructure, police, etc. for the community, which increases land value], or of the inhabitants of some particular place, enables them to pay so much more than its real value for the ground which they build their houses upon. . . . [The community creates land value.] Nothing can be more reasonable than that a fund [tax], which owes its existence to the good government of the state [good infrastructure, police, etc.] should be taxed peculiarly, or should contribute something more than the greater part of other funds, towards the support of that government.[198] [Nothing is more reasonable than taxing land exclusively and the land tax should be the greatest portion of all government tax receipts.]

Smith believed the LVT was the most socially beneficial and economically efficient tax because squatter income is unearned. In Smith's own words:

Both ground-rents and the ordinary rent of land are a species of revenue which the owner, in many cases, enjoys without any care or attention of his own. [Squatters suck wealth from an economy without adding any value to it.] Though a part of this revenue should be taken from him in order to defray the expenses of the state, no discouragement will thereby be given to any sort of industry. [A land tax should be extracted from squatters to fund the government; and a land tax does not disrupt any productive business activity.] The annual produce of the land and labour of the society, the real wealth and revenue of the great body of the people, might be the same after such a tax as before. [Land taxes do not

---

198 Smith, Adam. (1776). The Wealth of Nations. Book V, Chapter 2, para 76.

diminish the output and income of any productive enterprise.]
Ground-rents and the ordinary rent of land are, therefore, perhaps, the species of
revenue which can best bear to have a peculiar tax imposed upon them. ***[Land***
taxes are the most economically efficient form of taxation.]

Smith confirmed that separating land taxes from property taxes is easy; so, there's no rational reason for anybody (except squatters) to oppose it. In Smith's own words:

The contrivers of taxes have, probably, found some difficulty in ascertaining what part of the rent ought to be considered as ground-rent ***[land tax]****, and what part ought to be considered as building-rent* ***[property tax]****. It should not, however, seem very difficult to distinguish those two parts of the rent from one another.*[199]

A Sovereign Wealth Fund for Every Country

The LVT Gives Every Country a Large Sovereign Wealth Fund. Countries like Norway, Saudi Arabia, Qatar, Kuwait, the Emirate of Abu Dhabi, and the U.S. State of Alaska are envied around the world for their oil wealth. Their natural resource wealth gives them the ability to create massive sovereign wealth funds. Those governments provide their citizens with a higher quality of life than most Americans today can imagine. But land is a natural resource, too, and there's no reason why it can't be managed to create a sovereign wealth fund just like oil. In fact, if land was managed like oil in those countries—as a public natural resource—it would generate many times more wealth for every nation's citizens than oil because land is far more abundant and economically valuable *in every country.*

The LVT Would Eliminate Poverty Virtually Overnight. Given that land is created by Nature, it's immovable, and it cannot be significantly increased through any human effort, it's even more logical to classify land as a public natural resource than oil. Any country that adopts the LVT, manages their land as a public natural resource, and

---

199 Smith, Adam. (1776). The Wealth of Nations. Book V, Chapter 2, para 77.

creates an LVT-based sovereign wealth fund to provide meaningful social stability services to their citizens would eliminate all poverty in their countries virtually overnight.[200]

What Does a Competent Government Look Like? Only *0.5%* of Norwegian children are born into poverty.[201] In other words, 1 out of every 200 Norwegian kids is born into poverty. Norwegian politicians have managed their natural resources wisely and stockpiled over $1 trillion of wealth in their sovereign wealth fund.[202] They consume only 4% of that wealth per year to provide high-quality healthcare, education, a stable economy, and enriching cultural programs to the Norwegian population; and 4% is *far below* the rate of return on Norway's sovereign wealth fund, which is due to wise investment management. This is why Norway's economy produces broad-based wealth, *much higher* median quality of life, social stability and economic sustainability without sentencing their citizens to life-long debt slavery.

Which Government is More Competent? In contrast to Norway, approximately 32% of American children are born into poverty; i.e., an average of *64* out of every 200 American kids is born into poverty (64 times higher than Norway).[203] The U.S. Government has a $22 trillion debt, a $600 billion annual military budget, a nearly $1 trillion annual budget deficit, obscenely expensive and wasteful healthcare and education systems that are financially crushing the American population, the most indebted population in human history, the largest per-capita prison population on Earth, and one of the highest rates of poverty in the developed world. Which government do you think is more competent and has the most rational economic and political systems?[204]

---

200 Keep in mind that all those countries have non-citizen immigrants living in poverty, but a government is not responsible for the citizens of other countries. So, we don't include non-citizen immigrants in this analysis.

201 7 Important Facts to Know About Poverty in Norway. (2017, April 23). Retrieved May 2, 2018, from Borgenproject.org/7-facts-poverty-in-norway

202 Tappe, A. (2018, February 28). Norway's sovereign-wealth fund raked in a $131 billion return last year — here's why. Marketwatch.com/story/norways-sovereign-wealth-fund-rakes-in-131-billion-return-thanks-to-2017-stock-rally-2018-02-27

203 NCCP | Child Poverty. (n.d.). Retrieved May 2, 2018, from Nccp.org/topics/childpoverty.html. 32% derived from the high (43%) and low (21%) estimate provided by the NCCP.

204 To see a head-to-head data comparison between Norway and the U.S. and many other Continues on next page.

"But Nationalizing Land is Communism!" No matter how many real-world facts you present to some people, their ideological biases might prevent them from grasping the principles in this book. As you share this book and these ideas with others in your community, you might encounter people who falsely equate these principles with collectivism. For those people, be sure to emphasize the following points:

- The purpose of *human* economies and *human* societies is to provide the highest quality of life for the largest number of *humans*, not make a tiny number of corporate mercenaries super-rich and politically powerful while everybody else slides into poverty and political impotence.
- Land squatters destroy wealth, they destabilize economies, they do not add any value to any economy, and they suck wealth from everybody else. In fact, this is why Adam Smith had no respect for "the owner of the ground-rent [land squatter], who acts always as a monopolist. . . ."[205]
- The LVT re-aligns the incentives within an economy to reward *real* wealth creators while discouraging phantom wealth speculation by squatters.
- The LVT ensures that natural, financial, and human resources are deployed to the most economically efficient activities, which increases employment and investment opportunities throughout the economy, maximizes the employment rate, and fundamentally aligns all private economic activity with the needs of society.
- The LVT eliminates price and monetary inflation caused by speculative real estate bubbles, which preserves wealth and purchasing power for the general public so that they can buy the goods and services produced by real wealth creators.
- Natural resources are Nature's gift to the citizens of a nation, which no human or corporation ever creates; thus, no human or

---

countries, see: Eanfar, F. (2018) Global Governance Scorecard.
205 Smith, Adam. (1776). The Wealth of Nations. Book V, Chapter 2, para 74.

corporation has any right to *own* them.

- Real estate *developers* (not squatters), factory owners, and all other business owners that *develop* or use land for *productive* purposes that generate *real value* and *ongoing income* don't have anything to fear from the LVT. In fact, the single, simple LVT would be much less than all the combined taxes they're already paying today. That means the LVT would enable them to keep even more of their wealth than they do today. So, unproductive land squatters are the only group in any society that has a rational incentive to oppose the LVT.

- The LVT is a single tax, which means it would replace all other taxes. So, nobody can claim it adds to anybody's total tax burden (except for squatters).

- Natural resource extraction corporations can still make significant profits selling *their services* without owning and monopolizing land—Nature's gift to humanity.

- Managing natural resources as public resources has nothing to do with communism just like a government ensuring that its citizens have clean air and water is not communism.

- Air, water, land, and all natural resources are Nature's gift to humanity. It's time to start recognizing and institutionalizing that reality.

What About Other Natural Resources? If we're being intellectually honest and logically consistent, the LVT *should* be applied to all depletable, non-renewable natural resources. Additionally, the cause of most human rights abuses around the world is corrupt, private natural resource extraction companies that exploit and abuse their workers and keep virtually all the profits for themselves. Then they use those profits to financially capture politicians, which perpetuates the cycle of corruption that keeps their people in poverty.

However, compared to the total value of land, the value of all U.S. commodity production *combined* would only generate less than 1% of the annual LVT revenue. So, despite my sadness for the exploitation of workers in the developing world, it's currently not wise to fight all the special interest groups in the depletable commodities extraction industries until after the LVT is implemented for land.

The LVT Complements Cryptocurrencies

The LVT Eliminates the "Tax Collection" Argument Against Cryptocurrencies. With the LVT, politicians would have no legitimate excuse to block citizens from using cryptocurrencies, which are essential to human rights for many reasons. Anybody who understands how cryptographically secure assets are transmitted with cryptographically secure network protocols knows that nothing on Earth can stop the proliferation of cryptocurrencies, except for a worldwide calamity that physically destroys the Internet. Since the Internet is designed to survive a nuclear holocaust and billions of humans depend on the Internet every day, it's safe to assume that cryptocurrencies will survive all attacks by politicians and governments.

 The LVT Prevents Mass Incarceration & Murder. We can rationally expect that misguided politicians in many dysfunctional governments will use "tax collection" as an excuse to violently oppress, incarcerate, and murder large human populations. Why? Because mass resistance to tax tyranny is guaranteed to increase as politicians continue to doom their citizens to life-long debt slavery and destroy their economies by pandering to the expanding global corporatocracy. This will create many geopolitical consequences, which will also adversely impact relatively functional governments. This will increase the demand for well-designed cryptocurrencies, creating a vicious cycle of cryptocurrency demand and violent government oppression against cryptocurrency users.

 The LVT Prevents the Destruction of Human Civilization. If you truly understand the long-term implications of the preceding points, then you will understand why the LVT is the only way to prevent the destruction of human civilization as we know it today. This is not fear-mongering; this is the *predictable, inevitable and unstoppable* outcome of the convergence of technology, economics, geopolitics, and an expanding worldwide population of economic refugees. Respectfully, to all the humans on Earth today who are still asleep, please wake up!

Will the LVT Really Work?

This is a Very Brief Summary of the LVT. Before we continue, let's be clear about an important point: I could write an entire book about the LVT alone, including all the logistical and socioeconomic steps to get from the status quo to an LVT-based economy. However, this is not a book about the LVT; this book is intended to provide an overview of the most significant problems plaguing humanity today and the solutions that the Gini Foundation is building and supporting. So, please don't assume the LVT can be switched on overnight. Like any large-scale system, switching between the status quo and the LVT requires substantial time and effort. With that in mind, let's briefly summarize how the LVT would work in the real-world.

How Do We Measure Land Value? Land assessors measure the value of any parcel of land by two primary factors: (1) total productive output (pro rata GDP) that is occurring on the land and/or that *can* occur on the land based on nearby industrial activities; and (2) the attractiveness of the amenities associated with the land (nice climate, beaches, good schools, etc.). Those two variables alone account for between 80-90% of all land values; and pro rata GDP is dominant in determining the total value of land. So, land values in each city have a strong positive correlation with the city's contribution to total national GDP, which is why the table below includes the GDP figures as a reference point.[206]

How Much LVT Could Be Collected? Any alternative tax system that doesn't generate enough tax revenue to fund *all* local, state, and federal budgets would give politicians the pretext to add more taxes on top of the LVT. Fortunately, the value of land is so enormous that funding all levels of government from the LVT alone is easier and far less burdensome than the onslaught of taxes Americans pay today. The table below illustrates how the LVT works.

---

206 In this specific case, the 2018 US Land Value was derived from the latest 2009 estimate provided by the U.S. Bureau of Economic Analysis. The GDP growth rate between 2009—2018 was calculated and applied to the land value growth rate. This is reasonable because land values correlate strongly with GDP. See also: Larson, W. (2015). New Estimates of Value of Land of the United States. US Bureau of Economic Analysis.

How Much LVT Could Be Collected?

Existing Economic Factors	Amount $ Billions	
2009 US GDP	14,400	
2018 US GDP	19,965	
2009 US Land Value	23,000	
2018 US Land Value	31,889	
Federal Taxes (2018)	3,654	
State Taxes (2018)	1,096	
Local Taxes (2018)	365	
2018 Total Tax Collected (Fed/State/Local)	5,116	
Total Tax Collected % GDP (2018 Est)	26%	
Avg Current "Broken Tax" Rate	32%	
LVT Results		
Public Land Trust Fund Est.	5,116	= All F/S/L Taxes
Minimum US LVT Rate	11.5%	= Fed Taxes
Avg US LVT Rate	16.0%	= 100% F/S/L Taxes
Maximum US LVT Rate	20.0%	= 125% F/S/L Taxes
Avg LVT Rate vs. Broken Tax Rate	-50%	

Source: GiniFoundation.org

How Does the LVT Compare to the Current "Broken Tax" System? In the table above, all the values above the "LVT Results" line are actual and derived tax, GDP, and land value figures provided by the USG. All the values below the "LVT Results" line are tax rate tiers calculated to meet the indicated tax collection thresholds in the far-right column. The following is a brief explanation of what the values in the table mean.

- **Avg Current 'Broken Tax' Rate: 32%.** This is the current actual average aggregate tax rate for all federal, state, and local (F/S/L) taxes that Americans pay after all deductions, subsidies, and tax breaks.[207]
- **Public Land Trust Fund Est: $5.1 Trillion.** The arrows between this item and the "Avg US LVT Rate: 16%" indicate that the Public Land Trust Fund reaches 100% parity with the

---

207 "A Comparison of the Tax Burden on Labor in the OECD, 2016." Tax Foundation. Taxfoundation.org/comparison-tax-burden-labor-oecd-2016

actual 2018 taxes collected from *all* federal, state, and local governments. Thus, the average LVT rate of 16% is all that is necessary to fund all levels of government in the United States.

- **Minimum US LVT Rate: 11.5%.** This is the minimum LVT tax rate required to collect the same amount of taxes that the federal government collects today ($3.65 trillion in 2018).

- **Avg US LVT Rate: 16%.** This is the average LVT tax rate required to collect the same amount of taxes that *all* levels of government (federal, state, local) collect today (approx. $5.1 trillion).[208] State and local governments would have discretion to adjust the rate in response to natural local land value fluctuations, but 16% is the average nationwide LVT rate required to satisfy the budget requirements for all levels of government for the entire country.

- **Maximum US LVT Rate: 20%.** This is the maximum LVT rate ceiling. It is intended to give the federal government emergency authority to collect up to 125% of all normal federal, state, and local taxes to provide resources for a temporary natural disaster or a *legitimate* war that has been democratically approved by the citizens. In 2018, a 20% LVT would collect $6.4 trillion, which is 125% of the normal $5.1 trillion expected to be collected by all levels of government.

- **Avg LVT Rate vs. Broken Tax Rate: 50% Reduction.** The average LVT rate would only need to be 16% to achieve the same level of tax collection as the average 32% Broken Tax system rate today. This means the total tax burden that Americans would have with a single LVT would be 50% less than their total tax burden today.[209] This is possible because the LVT spreads the total tax burden across a much larger and more valuable land tax base and it eliminates USD trillions of tax

---

208 The total $5.1 trillion tax is allocated across each level of government as: 60% federal, 30% state, and 10% local. See: Tax Policy Center. (2016). Taxpolicycenter.org/briefing-book/what-breakdown-tax-revenues-among-federal-state-and-local-governments

209 Of course, in an LVT-based economy, the elimination of artificial inflation would reduce the price of everything. Thus, the price of land, cost of all government services, corresponding taxes, and all pre-LVT figures in the preceding table would decrease proportionately.

scams and unnecessary costs that plague the Broken Tax system today.

A Summary of LVT Benefits. For a single, simple 16% LVT expense, what do Americans get in return? The LVT:

- Eliminates all other taxes.
- Eliminates the IRS, its $11.5 billion budget, and the IRS' global police state tyranny.
- Eliminates the $1 trillion in wasted tax compliance costs.
- Eliminates the $450 billion taxes lost from illegal tax evasion.
- Eliminates all the Ponzi money, pyramid debt, and catastrophic market bubbles and crashes.
- Eliminates all special interest loopholes, corporate tax scams, shifting income to offshore tax havens, etc.
- Eliminates all urban sprawl because there would no longer be an economic incentive to waste any land.
- Eliminates the "cryptocurrencies make tax collection too hard" excuse.
- Enables the economy to operate at full employment because land and capital would be unlocked and maximally available to the *real* economy. Specifically, squatters would have strong economic incentives to put their wealth to work in *productive* enterprises, not unproductive land.
- Eliminates all the toxic effects from squatter-induced currency debasement, social class conflicts, and other socioeconomic maladies.
- Enables a gradual restoration of middle-class wealth and political power, which is the key to political and economic stability in every country.
- And many other benefits. . . .

The LVT Could Decrease Over Time. The prospect of having a single average 16% LVT is already inspiring, but that's simply the average nationwide rate that is initially required to support the obscenely bloated and dysfunctional federal government that we have today.

However, as all the LVT benefits percolate throughout the economy, it would inspire and reinforce many other important economic and political reforms, which would impose more discipline and transparency on all levels of government. That would squeeze out most of the waste, fraud, and abuse within the government, which would decrease the size of government. That would reduce the budget and corresponding LVT required to fund the government; thus, we could reasonably expect the initial average 16% LVT to be even lower over time.

The LVT is Not "Land Reform". Anybody who equates the LVT with "land reform" doesn't understand what the LVT is. "Land reform" is often associated with developing countries that suffer from periodic revolutions and cycles of land concentration, land reform, more concentration, more reform, etc. Land reform usually doesn't work because the government simply confiscates land from large landowners and then redistributes it to a bunch of smaller landowners, but that doesn't take into account how wealthy, politically connected people inevitably seek to consolidate the land again and again. Given that land is a natural resource and *should not be owned by individual humans or corporations at all*, obviously the concept of "land reform" has nothing to do with the LVT.

Is the LVT a Flat Tax On Everybody? If So, Isn't That Regressive? No, because only people who are extracting ongoing value from land (e.g., business owners and homeowners) will have an economic incentive to pay the LVT. Everybody else who rents property (home, apartment, building) from another property owner (landlord) contributes indirectly to the Public Land Trust by paying a market-based rent to their landlord just like they do today. Then, the landlord is responsible for paying the LVT. Thus, with the LVT, everybody contributes. For anybody with enough wealth to own property, the single LVT is a bargain. So, the LVT is not regressive at all.

The LVT Rate Is Market-Based. The federal government would set a minimum LVT rate (e.g., 11.5%) to ensure that the federal budget is covered, but anything above the minimum rate would be at the discretion of local municipal authorities. In fact, the average 16% LVT rate presented here is substantially higher than the typical land leaseholder would pay. The median LVT would be closer to 13%

because in over 99.5% of American cities the LVT would be substantially lower than 16%. This is because U.S. GDP and corresponding land values are highly concentrated in high-population-density metro areas like New York City, Los Angeles, Chicago, Houston, Miami, etc. High population density areas have stronger land demand than low-density areas; thus, their land prices will naturally be relatively higher than prices in low-density areas.

The LVT Produces a Fair & Constitutionally Compliant Tax Distribution. High-density metro areas already pay a substantially higher proportion of the total federal tax burden, which is what the U.S. Founders intended with the Apportionment Clause in the U.S. Constitution. The LVT would result in the same overall national distribution of the federal tax burden, but within each state, the LVT's much broader land tax base results in a much lower average rate for taxpayers.

How Would the LVT Impact the Gini Team? Chapter three started with a section entitled, "Self Interest vs. Humanity's Interest." The principles in that section are crucial to remember in every public policy discussion. In this case, virtually everybody in the Gini Foundation currently owns real estate. So, we have thought deeply about how the LVT would impact our own personal wealth. Without hesitation, we are 100% confident that the wealth of humanity (including us) would be substantially increased by the LVT and all the solutions presented in this book.

The LVT Is Only Half the Solution.

The LVT is the solution to substantially eliminate poverty and many other socioeconomic problems in every country. However, giving politicians large sovereign wealth funds without also ensuring there is sufficient institutional integrity to hold them accountable for their budgetary expenditures will simply lead to more corruption, more wasted resources, and more tyranny. That's where Gini comes into the picture. Now, we're ready to discuss why the Gini ecosystem we are building with many like-minded humans around the world is the most effective way to protect human rights, maximize the broad wealth-generating potential of real-world commerce, and achieve the real-world humanitarian goals that we've discussed so far.

Key Points

- **Squatters Destroy Economies & Communities.** The endless monetary inflation, currency debasement, financial crises, structural unemployment, destruction of the middle class, rapidly expanding poverty, and the vast majority of socioeconomic problems that humanity suffers from today are primarily caused by land squatters and the banks that create mountains of Ponzi pyramid debt by over-leveraging real estate-based collateral throughout the global economy.

- **The Land Value Tax Is the Perfect Tax.** Adam Smith, Henry George, and numerous Nobel Prize-winning economists (including Paul Samuelson, Milton Friedman, Joseph Stiglitz and William Vickrey) all agree: The LVT is by far the simplest, most equitable, most economically efficient and rational form of taxation ever conceived. This book succinctly explains why this is true within the real-world context of the modern global economy.

- **There Is No Rational or Legitimate Reason to Reject the LVT.** Anybody who rejects the LVT with vague concerns about "logistical challenges" or claims the LVT is based on communism is either a land squatter, paid by land squatters, or has no idea what they're talking about. The primary reason humanity is tortured by a blizzard of taxes today is because there has never been an energetic, well-funded campaign to push back against the squatter lobby and to help the general public understand the amazing benefits of the LVT. This is why Gini is creating the online Gini School of Economics, which will help humanity learn about Gini's unique economic philosophy, including our strong support for the LVT.

- Chapter 6 -
Gini: Blockchain with a Soul

"A patriot must always be ready to defend
his country against his government."
— Edward Abbey

———————⚜———————

Cryptocurrencies have the potential to revolutionize human civilization, but only if they're designed and implemented with an accurate understanding of the real-world problems discussed throughout this book. Before we summarize Gini's technical architecture and features, it's useful to briefly summarize the reason the Gini Foundation exists and how we perceive cryptocurrencies within the context of the human rights concerns discussed in the previous chapters.

Gini is Focused on Sustainable *Human* Markets. The Gini Foundation was founded on a simple but important principle: Sustainable *human markets* should benefit *actual* humans. We emphasis "sustainable" and "human" because all other cryptocurrency projects today are dominated by *whales,* which may or may not be human, but their impact is destructive to *actual* human markets. Additionally, other cryptocurrency projects substantially ignore the technical, economic, geopolitical, sociological, and humanitarian consequences of *non-human* artificial intelligence; the overgrowth of gigantic, *non-human* corporations; and incompetent, corrupt and *inhumane* politicians that are killing capitalism and democracy worldwide today.

Building the Technology is Not Enough. The leadership teams of other cryptocurrency projects assume that all they have to do is build a soul-less set of blockchain tools and then those tools will be automatically used for the good of humanity. The Gini team wishes that were true, but we have enough real-world technical, business, economics, and geopolitical experience to know it's absolutely not true. This is another reason we're building the Gini School of Economics.

Justice, Equity and Fairness as a Service. Sometimes we refer to the Gini Platform as "Justice, Equity, and Fairness as a Service" (JEFS) because these principles are the essence of what every democratic socioeconomic system should achieve. The power of blockchain-based technologies to solve many institutional integrity problems is awe-inspiring. However, without clarity of purpose and a relentless focus on justice, equity, and fairness at the core of every technical design decision, it's too easy for technical systems to be co-opted by self-serving politicians and corporations.

Ideology Drives All Cryptocurrency Projects. Some people assume scientists and engineers involved in cryptocurrency projects are immune to ideology. Not true. In fact, ideology is at the heart of every cryptocurrency.[210] Science, Technology, Engineering, and Math (STEM) skills alone cannot prepare engineers to understand and deal with real-world problems associated with the economic survival of humans in an age of artificial intelligence, generational poverty, wealth and income disparities, war and geopolitical conflicts, the balance of geopolitical power between nations, the politics of environmental sustainability, the politics of taxes and industry regulations, and a long list of socioeconomic problems. Because of this, most younger technical people naturally gravitate to the simplest ideology of all—Libertarianism—because it's easy to say, "Government, leave me alone!" and then go back to coding. Unfortunately, human civilization is more complicated than that.

The Gini Foundation's Roots. The Gini team has decades of technical experience, but we also have real-world experience in business, economics, geopolitics, and we are well-educated in classical Political Economy, including works by Adam Smith, J.S. Mill, Montesquieu, the Federalist Papers (James Madison, Alexander Hamilton, John Jay), Thomas Jefferson, John Locke, Thomas Hobbes, among others. These Enlightenment Era Political Economy philosophers illuminated the path

---

210 Even Alan "The Maestro" Greenspan admitted his puritanical libertarian ideology was flawed in his congressional testimony about why his ideology blinded him to the problems in the U.S. economy. See: Leonhardt, D. (2008, October 23). Greenspan's Mea Culpa. Economix.blogs.nytimes.com/2008/10/23/greenspans-mea-culpa
Also Watch his testimony here: Youtu.be/R5lZPWNFizQ?t=333

to the American Revolution and the subsequent worldwide explosion of human liberty and prosperity. The Gini Foundation is founded on *those principles*, which emphasize *human value* and *human freedom*. In contrast, all the major cryptocurrency projects today generally focus on the principle of *transaction freedom*, but they ignore many other aspects of human existence that actually produce *genuine and sustainable* economic, political, and *human* freedom in the real world.

What Does the Name "Gini" Mean? The name "Gini" is based on an important principle in the field of Economics called the Gini Index. The Gini Index is a well-known statistical tool to measure the dispersion of any quantity in any domain, but it's most famously associated with measuring the distribution of wealth within human populations and nations. Every population has a Gini Index value that is somewhere between two extremes: A Gini Index of zero (or 0%) means everyone has the same amount of wealth. In contrast, a Gini Index of 1 (or 100%) means only one person has all the wealth, and everybody else has none. Each country's Gini Index is strongly correlated with many important economic and social consequences. For example, Transparency International observed:

> *This year's results highlight the connection between corruption and inequality, which feed off each other to create a vicious circle between corruption, unequal distribution of power in society, and unequal distribution of wealth.*[211,212]

The Mathematical Meaning of Gini. The Gini coefficient (or Gini Index or Gini ratio) is a statistical summary of the Lorenz Curve, which measures inequality in a population.[213] Mathematically, it's often described as the "relative mean difference"; i.e., the average of the difference between every possible pair of values in a given distribution, divided by the average value. The Gini Index formula is below.

---

211 Corruption Perceptions Index 2016.
Transparency.org/news/feature/corruption_perceptions_index_2016
212 Corruption Perceptions Index 2016.
Transparency.org/news/feature/corruption_perceptions_index_2016
213 Dixon et al. 1987; Damgaard and Weiner 2000.

$$G = \frac{\sum_{i=1}^{n} \sum_{j=1}^{n} |x_i - x_j|}{2n^2 \mu}$$

The Humanitarian Meaning of Gini. The consequences of high Gini Indexes are profound. Whenever a country's Gini Index is high, democracy inevitably suffers. When democracy suffers, the Gini Index tends to increase until economic oppression ignites violent social and political revolutions. This is not about ideology; it's about basic human nature. When large human populations are oppressed by gigantic corporations and self-serving politicians, they will naturally resist that oppression.

The Gini Index is Humanity's Early Warning System. The Gini Index represents an early-warning system to help economic policymakers make adjustments to release the socioeconomic pressures in their countries before violent revolutions start. This is the essence of what the Gini Foundation is working to accomplish by providing humanity with the tools that are essential to protect human rights and maximize the broad wealth-generating potential of real-world commerce.

The Technician's Response to Human Suffering. Over many years of technical development, we have noticed that many engineers have not read the classical political economists. So, they often don't understand economics and geopolitics beyond the basic talking points that they are spoon-fed in shallow Economics classes or the Wall Street-dominated financial news media. They may develop strong opinions based on what they see or hear, but they tend to shy away from topics like socioeconomics and geopolitics because they don't feel confident discussing topics that can't be cleanly resolved with an elegant algorithm.

Code is Not Law. The complexity of economic and geopolitical systems often causes software engineers to over-simplify the world by declaring "code is law!" without really understanding how their unconscious ideologies infiltrate their source code. This can prevent them from accurately recognizing cause-and-effect relationships within socioeconomic and geopolitical systems, which prevents them from using their technical talents to cure the tragic injustices that plague humanity today.

Libertarianism Has No Meaningful Response to A.I. & Transnational Cannibals. The Gini team has a deep understanding of the libertarian ideology that guides virtually all other cryptocurrency projects today because everybody on the Gini team was a hardcore libertarian for most of their lives. But over time, we realized that Libertarianism has no meaningful response to a world in which artificial intelligence is gobbling up virtually every human job.[214] Libertarianism has no meaningful response to a world in which politicians reward gigantic transnational corporations with de facto monopolies in exchange for cooperation in their rapidly expanding surveillance state.[215] In this *real world* (not the utopia of zero government intervention), humans and small- to medium-sized companies have virtually no chance to compete against A.I. and "Transnational Cannibals" . . . unless something changes.[216]

Technology is Meaningless & Self-Destructive without Clarity of Purpose. Technology is meaningless—or worse, it's destructive to humanity—to the extent that it is not developed and implemented by humans with properly aligned incentives, a deep understanding of how the technology can be deployed to solve real-world humanitarian problems, and the courage to resist the attacks of opposing groups that are invested in the status quo. The Gini technical whitepaper (available at GiniFoundation.org) and the chapters in Part 2 of this book focus on three core areas: Monetary Policy, Technology, and Project Governance. Unlike other cryptocurrency whitepapers, we spend a significant amount of time discussing Gini's unique monetary system because monetary policy is what impacts stakeholders in every fiat *and* crypto economy more than any other factor.

Technology is Meaningless & Self-Destructive if it Ignores Real-World Realities. To be clear, the Gini Platform is being built with the most advanced, provably secure, and robust blockchain technology on Earth today, which is covered later in the Technology chapter. In fact, several Gini team members have been involved with cryptographic

---

214 The Collapse of the Human Labor Force: Eanfar.org/collapse-human-labor-force

215 Proof of Google Censorship/Path to Freedom: Eanfar.org/proof-google-censorship-path-freedom

216 Transnational Economic Cannibals: Eanfar.org/transnational-economic-cannibals

technologies for over a decade before Satoshi Nakamoto launched Bitcoin in early 2009; so, we have substantial technical skills and insights to fuel our long-term technical development. However, it's even more important that we have a deep understanding of real-world Political Economy (not only textbook Economics) because no cryptocurrency technology will achieve anything significant if it violates or ignores the fundamental dynamics of human nature, sociology, economics, and geopolitics.

Irrational Idealism vs. Rational Humanitarian Crisis Prevention. When animals are faced with an existential threat, they resist and fight until that threat no longer exists. They don't ponder whether they are being too idealistic or not. They don't care what other self-serving and self-destructive animals might say. They fight and resist because the alternative is death and/or the destruction of their ecosystem. In human societies, there is nothing irrational or unreasonably idealistic about identifying existential and ecosystem threats and then investing our time, talents, and resources into preventing those threats from destroying us. In fact, the *irrational choice* is doing nothing when we know there is an alternative path.

The Alternative Path. The economic and political systems that exist on Earth today are socially constructed frameworks dominated by a tiny number of humans who benefit from the status quo. These frameworks are not permanent, immutable features of human existence; they are fabricated by the collusion between the largest shareholders of gigantic corporations and politicians that have self-serving incentives to preserve the status quo. The status quo exists because we, as a species and as human societies, have relinquished control of our economic and political power to gigantic corporations and politicians without demanding meaningful accountability and an equitable share of the economic pie in return. The status quo is killing markets, capitalism, democracy, and human liberty on Earth today. We have a choice: Gini is an alternative path forward; and we hope you will choose to join us on this journey.

The Evolution of Cryptocurrencies

Early Pioneers that Made Cryptocurrencies Possible. The basic principles of a blockchain have their roots in a Computer Science concept called "linked lists," which were developed in 1955 by several research scientists at the Rand Corporation. Linked lists became more prominent after they were integrated as a primary data structure in the LISP programming language, which was developed by John McCarthy in 1958 while he was a research fellow at MIT.[217,218] However, research into cryptocurrencies, in particular, began in 1983 with David Chaum's paper, "Blind Signatures for Untraceable Payments. Advances in Cryptology Proceedings of Crypto."[219] Five years later, Chaum, et al., explored the concepts of payment anonymity, prevention of the "double-spend" problem, and potential defenses against malicious network attackers in their paper, "Proceedings on Advances in Cryptology."[220]

The Advent of Proof-of-Work & HashCash. Another significant breakthrough came in 2002 when Adam Back invented HashCash, which used a pre-existing Proof-of-Work algorithm (previously invented by Cynthia Dwork and Moni Naor) to prevent email spam.[221,222] This was the last ingredient that Satoshi Nakamoto needed to build Bitcoin seven years later. Today, there are several thousand cryptocurrency *tokens*, but the vast majority of them are based on the Ethereum or Bitcoin blockchains, which means they suffer from the same limitations as those blockchains, as we will discuss later.

The Birth of Bitcoin. In late-2008, a pseudonymous person (or

---

217 McCarthy is also widely regarded as one of the founding fathers of artificial intelligence, which is a phrase he coined.

218 Roberts, Jacob (2016). Thinking Machines: The Search for Artificial Intelligence. Distillations. 2 (2): 14-23.

219 Chaum, D., 1983: Blind signatures for untraceable payments. Advances in Cryptology Proceedings of Crypto 82 (3): 199–203

220 Chaum, D., Fiat, A., Naor, M., 1988: Untraceable electronic cash. CRYPTO 88 Proceedings on Advances in Cryptology, pp. 319-327, Springer-Verlag New York, Inc. New York, NY.

221 Pricing via Processing / Combatting Junk Mail. Wisdom.weizmann.ac.il/~naor/PAPERS/pvp_abs.html

222 Back, A., 2002: Hashcash - A Denial of Service Counter-Measure. Hashcash.org/papers/hashcash.pdf

group) named Satoshi Nakamoto changed the world, but it would take years for most people to realize it. Shortly after the 2008 Financial Crisis, Nakamoto published a technical whitepaper entitled, "Bitcoin: A Peer-to-Peer Electronic Cash System," but virtually nobody read it. However, after Nakamoto converted the ideas in the whitepaper into a relatively functional blockchain software program in early 2009, a few cryptography-based discussion forum denizens started paying attention.

The Problems. The primary problems that Nakamoto was trying to solve with his new digital currency were timely and relevant to every American and most humans worldwide at the time because the shockwaves of the 2008 Financial Crisis were still shaking our planet. Gigantic *too-big-to-fail* banks, politicians and governments captured by the banks and other wealthy special interest groups, transnational cannibals with no meaningful loyalty to their communities or countries, a central bank that was (and still is) destroying the U.S. Dollar's value, levels of wealth concentration not seen since the Robber Barons of the late 19th and early 20th Centuries . . . these were all problems that deeply disturbed Nakamoto, which can be discerned from his writings.

Compulsory Institutional Trust Is the Core Problem. The preceding list of problems have a common factor: compulsory trust in corruptible, centralized institutions. We are forced to trust central banks to manage fiat money supplies, but human history over the past 1,200 years is a veritable graveyard of fiat currencies that were debased and ultimately destroyed by irresponsible and manifestly untrustworthy politicians and central bankers.[223] We are forced to trust commercial banks to keep depositors' funds safe, but their untrustworthy business models are inherently dependent upon a banking system that inevitably blows up systemically unsustainable debt bubbles, which destroys the wealth and liberty of their depositors when those bubbles pop. Then, these untrustworthy institutions exploit preventable crises to further concentrate their power over the general public, which destroys democracy and capitalism in many ways.

Central Banks Start Paying Attention to Blockchains. The

---

223 See "Fiat Currency Graveyard: A History of Monetary Folly": GiniFoundation.org/kb/fiat-currency-graveyard-a-history-of-monetary-folly

cryptocurrency revolution started quietly, grew slowly, and was completely invisible to all but a relatively small number of *cypherpunks* (cryptography enthusiasts) until around 2016. Then, something happened that transformed Nakamoto's invention—Bitcoin—into a household name. The Bank of England announced in January 2016 that it was exploring the possibility of using distributed ledger technology (blockchains) for the UK's bank settlement system. Soon after, several other major central banks made similar announcements, including the European Central Bank and the Bank of Japan, while the U.S. Federal Reserve made slightly more ambiguous but substantive statements about the technology.[224]

Commercial Banks and Big Corporations Jump on the Bandwagon. In April 2016, IBM and over a dozen of the largest banks in the world announced they were forming a consortium with the *Terminator-esque* name "R3," which is dedicated to building distributed ledger solutions for banks and large corporations.[225] Given the agendas of all the corporations involved, predictably, R3's definition of "distributed ledger" bears virtually no resemblance to Bitcoin or any other genuinely decentralized, distributed ledger.

In fact, all transactions on R3's networks are centralized and controlled by the consortium; they're directly monitored by the U.S. Government; transactions are not shared across the network by default; they do not protect the privacy of end-users from the consortium or potentially corrupt politicians; and they're designed to maximize the profitability of the consortium members, which has nothing to do with human rights or the open source spirit they claim to be supporting.[226]

Cryptocurrencies Explode into the Public Consciousness. In December of 2017, the price of Bitcoin rocketed to nearly $20,000 per unit and several other cryptocurrencies experienced even faster price rises between late-2017 and early 2018. These events inspired millions of

---

224 Bank of England to Explore Distributed Ledger Tech for Settlement. (2016, January 28). Coindesk. Coindesk.com/bank-england-distributed-ledger-settlement

225 R3 Announces New Distributed Ledger Technology Corda. (2016, April 5). Coindesk. Coindesk.com/r3cev-blockchain-regulated-businesses

226 See also: "The Emerging Crypto Banking Cartel": Eanfar.org/the-emerging-crypto-banking-cartel and "Blockchain Patent War Coming": GiniFoundation.org/kb/blockchain-patent-war-coming

humans to enter the crypto markets with dreams of becoming crypto millionaires overnight. They also inspired thousands of entrepreneurs to jump into the game, including many fraudsters and scammers.

Here Come the Regulators. Naturally, the influx of scammers has attracted the attention of government regulators. In general, this is a positive development for the cryptocurrency industry as long as regulators don't use the scammers as an excuse to try to shut down the entire industry. Given the collusive history of banks and politicians, we know the banks are already trying to use their lobbying power to perversely claim that they're the only entities that should be allowed to use cryptocurrency technologies. However, that tactic will not work in the long-run because there are always ways to circumvent Internet censorship. So, the primary question is: Will governments resort to violence and accelerate the destruction of our human rights to give banks an unfair advantage in the crypto world the same way they do in the fiat world? We will answer that question from several perspectives in the following chapters.

Cryptocurrency Value

Despite the recent spike in attention given to cryptocurrencies, many humans on Earth still don't really understand why cryptocurrencies have real-world value beyond the speculative hype they might see in the media. This section is intended to briefly summarize why cryptocurrencies have tremendous value relative to fiat currencies and other assets. So, let's start with the basic concept of how value flows through an economy.

What is Value? As explained in deeper detail in my previous book, *Broken Capitalism: This Is How We Fix It*, "value" is an invisible substance that permeates every aspect of human existence. The flow of value in human relationships and societies enables humanity to innovate, build amazing technologies, construct the tallest skyscrapers, and experience the most intimate love between romantic partners. We exchange value with others when we buy their products and services, recognize their accomplishments, hug them at dinner parties, and engage in sensual intercourse. In short, value is what drives human civilization forward to achieve ever-greater heights in every dimension of human existence.

What is Money? Money is how we exchange value between humans, organizations, and automated nodes on computer networks. Money gives us the ability to buy and sell goods and services. Money represents the value that we see, exchange, and experience when we use tangible products and intangible intellectual property. Money is what enables human societies to create economic systems, which enables us to efficiently organize resources, accumulate wealth, and deploy our accumulated wealth in the form of capital investments that fuel the growth of companies and entrepreneurial dreams. In short, money gives us a more concrete way of measuring and managing the flow of value in human societies so that we can individually and collectively achieve the quality of life that we desire.

What is a Currency? Money and currencies are not synonymous. Money can be anything (coins, seashells, fiat and cryptocurrencies, apples, lollipops, baseball cards, etc.) used to exchange value between two entities. In contrast, currencies have a more rigid definition and purpose. Specifically, currencies must fulfill three fundamental requirements:

- **Medium of Exchange.** A currency is a *medium* through which we measure and exchange value with others. In this respect, it is functionally the same as money, but money becomes a currency when it embodies the other two characteristics below.

- **Store of Value.** A currency enables us to store value in it for later use. This is how wealth is accumulated for a rainy day or future investment. An example is gold coins: We can store value in them and then exchange that value later for other kinds of value that we need or desire.

- **Unit of Account.** When you buy and sell anything, how do you count the money you use to buy it? How do you count the profits and losses when managing a business? What do you print on receipts to record how much money (value) you have spent in exchange for the value received? Accounting systems of all kinds help us answer these questions. In fact, every accounting system depends on a unit of *account* to ensure that the units of value that are used in each transaction are mutually understood and mutually acceptable between counterparties.

Are Cryptocurrencies More Valuable than Gold? The global economy is valued at approximately $100 trillion (in 2018) and the amount of fraud is often estimated at between 10-20% of global economic activity. That means the tamper-proof benefit *alone* of a well-designed cryptocurrency can be estimated at up to $20 trillion, not including all the other benefits. Considering that the global gold market is valued at only about $7.5 trillion (2017 World Gold Council estimate), the tamper-proofing value of a well-designed cryptocurrency justifies an aggregate cryptocurrency market price that is *at least* $10 trillion greater than the $7.5 trillion global gold market.

Connecting Tangible Value to Crypto Value. Before the emergence of cryptocurrencies, all the members of the Gini Foundation team were *gold bugs* for all the usual reasons; so, we appreciate the unique historical role that precious metals have served in many economies throughout human history. In fact, the Gini Foundation is working with hundreds of gold, silver, and copper miners in several developing countries to enable them to securely transact with their customers all over the world on the Gini *Decentralized* Exchange without being exploited by gigantic mining corporations and distribution middlemen. This will enable them to create and preserve wealth directly within their own communities.

Is Gold *Better* than Cryptocurrencies? Some people think cryptocurrencies are inferior to gold because gold is a tangible material and cryptocurrencies are not. Respectfully, people who believe this don't truly understand the following important concepts.

- **"Value" in a modern global economy** is increasingly based on the fluid flow of *intangible* ideas, the *intangible* trustworthiness of cryptographically secure data and information systems, the *intangible* trustworthiness of incorruptible institutions based on *decentralized* political and economic power structures. . . . These are the essential foundations of virtually every *tangible* product and service in every modern economy. Gold has virtually no place, purpose, or relevance in any of these domains of modern economic activity; thus, it has virtually no place, purpose, or relevance in modern human civilization beyond its industrial, collectible, and jewelry-making value. *However*, precious metals are still widely *perceived* as a stable store of value; and that *perception* alone makes them valuable to many people.

- **Precious metals prices can be manipulated** by governments, central banks, and powerful economic actors far more easily than the price of cryptocurrencies *at scale;* i.e., when they are used for a significant percentage of an economy's total economic activity. This is because the supply of precious metals can be manipulated by powerful entities more easily than the supply of a well-designed cryptocurrency, provided it is based on a well-designed monetary policy. (This topic will be discussed in more detail in the next chapter.)

- **Gold encourages centralization of gold reserves.** The logistical, security, and insurance requirements to manage gold safely requires vault operators to spend USD millions on very expensive facilities and property insurance. This inevitably leads to high concentrations of gold in centralized vaulting facilities because only the largest and most well-funded corporations can afford to operate these facilities effectively. Centralization makes large gold stockpiles highly vulnerable to political events, government confiscation, and many forms of corporate manipulation and institutional corruption.

- **It's easy for persistent governments to confiscate your precious metals**, which has happened multiple times in multiple countries in the past with gold and silver.[227,228] In contrast, it's virtually impossible for any government to confiscate a privacy-assured cryptocurrency if you manage them properly.

- **Fiat currencies *and* precious metals can be counterfeited.**

---

227 Executive Order 6814. Presidency.ucsb.edu/node/208602
228 Executive Order 6102. Many sources here:
Wikipedia.org/w/index.php?title=Executive_Order_6102

Sophisticated scanning equipment and chemicals *can* be used to verify the authenticity of paper currency and precious metals units, but that can be an expensive, tedious, and time-consuming process. This makes fiat and precious metals impractical for frequent commercial transactions without depending on centralized authorities like banks, governments, and precious metals refineries to verify every transaction. That exposes humanity to all the problems discussed in the preceding chapters.

- **Cryptocurrencies are counterfeit-proof.** In contrast to fiat and precious metals currencies, it's impossible to counterfeit a well-designed, decentralized cryptocurrency that is stored on a cryptographically secure blockchain. In fact, this is one of the reasons why many governments today are considering the possibility of converting their fiat currencies into cryptocurrencies.

Government-Issued Digital Currencies Are Still Fiat Currencies. Even after governments convert all their physical fiat currencies to digital currencies, they will still be fiat currencies because they will still be vulnerable to *official counterfeiting* by politically-driven politicians and central banks. Yes, *money printing, quantitative easing, and fractional reserve lending* are all forms of *official counterfeiting*, regardless of what self-serving politicians and bankers may claim. A government's *legal monopoly on counterfeiting* based on its ability to coerce citizens at the barrel of a gun (or a long prison sentence) does not change this reality.

Government-Proof. As the comparative quality-of-governance data in the Global Governance Scorecard confirms, politicians in many countries today are running their countries into the ground.[229] They are manipulating their national currencies, incurring huge and unsustainable debts, and destroying their economies, which is destroying their societies. To cover up their economic policy mistakes, they're probably also planning to control the capital flows in their economies by implementing unjust *bank bail-ins* and other anti-democratic measures to

---

229 Eanfar, F. (2018). Global Governance Scorecard. Eanfar.org.

prevent their citizens from escaping the economic tyranny they've spawned with their short-sighted and broken economic policies.

Cryptocurrencies Can Protect Our Wealth & Quality of Life. To be clear, I'm not gratuitously attacking politicians or governments; what I'm describing is simply the objective reality of broken capitalism on Earth today, which is destroying our global economy, eviscerating the middle class in many countries, spawning increased tensions between countries, and thus, creating the conditions for World War III. A well-designed cryptocurrency can prevent rogue politicians and corporations from destroying our wealth, our economies, and the quality of our lives.

Price Stability. Although the price of many cryptocurrencies today is more volatile than many fiat currencies, certain cryptocurrencies are fundamentally designed to achieve more price stability as their markets grow. Gini, in particular, is designed with specific features to make its market price more stable than other cryptocurrencies (and many fiat currencies). As the demand for Gini grows, the inherent price-stabilizing mechanisms of the Gini Platform are expected to continue reducing the price volatility of the Gini currency over time.

Collapsing Fiat Currency Value. The chart below illustrates how the U.S. Dollar has lost over 98% of its value since the beginning of the 20th Century. That's not an accident: The beginning of the 20th Century is precisely the time when U.S. politicians began launching foreign wars to spread U.S. economic imperialism under the false flag of *spreading democracy*. The Iraq War is just the most recent example, but this has been a habitual vice of U.S. politicians and several other governments for generations. Contrary to the principle of consensual democratic governance, these wars force citizens to endure tragic costs in life and wealth, resulting in immoral debts that are thrust upon future generations. We have now reached a negative tipping point in human history and the cycle of wars and debts is choking the life out of humanity economically and socially. The Gini cryptocurrency can help stop this.

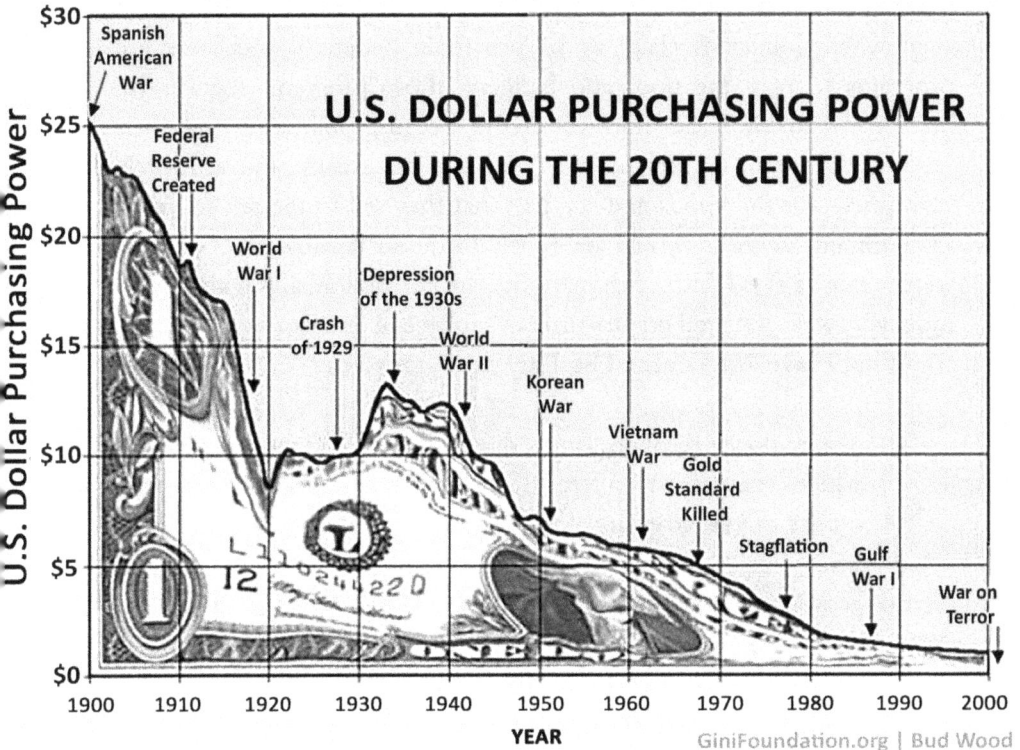

U.S. DOLLAR PURCHASING POWER DURING THE 20TH CENTURY

GiniFoundation.org | Bud Wood

Killing the USD and Our Wealth. The loss of over 98% of the USD's value is the *inevitable consequence* of how the fractional-reserve banking system works today *when it's clogged up with toxic speculative debt.* Visualize the mutilated USD bill in the chart above being stabbed with a butcher knife, slashed with a machete, hacked with a chainsaw, acid poured on its face, and brutalized like a domestic violence victim—that's the tragic crime that U.S. politicians have committed against the USD and American citizens since the beginning of the 20th Century. Now, U.S. politicians are dragging the USD into the fiat currency graveyard.[230]

The Collusion Between Governments and Corporations Destroys the Value of Our Money. We already know that politicians in many countries have abused their fiat currencies by funding immoral wars and bank bailouts, both of which are massive corporate welfare

---

230 See the currency graveyard here: "Fiat Currency Graveyard: A History of Monetary Folly". GiniFoundation.org/fiat-currency-graveyard-a-history-of-monetary-folly

programs for the military-industrial complex and gigantic banks. But even when politicians claim to be providing beneficial social welfare programs to help the poor and indigent, those programs metastasize into unsustainable fiscal monsters. Why? Because politicians give their favorite corporate donors no-bid contracts, tax-breaks, and artificially high prices for the goods and services that they sell to the government. Government services would be much more sustainable if politicians were acting in good faith, without corporate corruption, and with honest money. A well-designed cryptocurrency enforces honest money.

What Have We Learned So Far? Let's summarize:

- Value flows through every dimension of human civilization, which enables us to create amazing things and achieve the quality of life we desire.
- Money enables us to measure and manage the flow of value in economic transactions.
- Currencies ensure that we can efficiently and accurately keep track of the value we exchange with others.
- Politicians, gigantic banks and corporations have destroyed the value of fiat currencies, shackled us with unsustainable and asphyxiating debts, and are leading us to World War III.
- All these problems will inevitably get much worse as artificial intelligence, automated robots, the Internet of Things (IoT), ubiquitous government surveillance, and the hyper-concentration of political and economic power continue to erode our political freedom and the integrity of economies and human societies worldwide.
- A well-designed cryptocurrency cannot be manipulated and debased by banks, corporations, or governments.
- Cryptocurrencies provide all the benefits of fiat currencies (medium of exchange, store of value, unit of account), but they also have several additional benefits that prevent politicians and corporations from destroying our economies, our societies, our lives, and our planet.

Cryptocurrency Technologies

Currently, there are over 2,000 active *crypto tokens*, but nearly all of them are focused on niche applications, which means they're not suitable as currencies for real-world commerce. Additionally, none of the other cryptocurrencies have the combination of true privacy, scalability to thousands of transactions per second, a truly decentralized network topology, an integrated decentralized exchange to prevent cryptocurrency exchange monopolies, a truly sustainable monetary system, and substantial ecosystem stability mechanisms to incentivize mass adoption by consumers *and* merchants in real-world commerce.

In contrast, the Gini Platform is designed to deliver that combination of performance characteristics and protect citizens against the human rights abuses covered in previous chapters. However, before we discuss the Gini technology specifically, let's briefly summarize how blockchains work in general and some major flaws in the way existing cryptocurrencies are designed today within the context of protecting human rights.[231]

What is a Cryptocurrency? A well-designed cryptocurrency has all the essential features of any major currency: medium of exchange, store of value, unit of account. Additionally, because they're cryptographically secured on a blockchain, cryptocurrencies can provide a level of privacy and transaction integrity that cannot be achieved by other payment mechanisms like credit cards, PayPal, and bank cash deposits. Finally, like any other asset, a well-designed cryptocurrency has value because a community of humans believes it has value for their particular purposes and needs (i.e., *utility* value).

What is a Blockchain? The easiest way to understand the concept of a blockchain is to visualize a spreadsheet, which every computer on the Internet can see. But in this spreadsheet, all the transaction values are entered from the bottom up. After you insert a piece of information into one cell (block), somebody else can insert another block on top of it. Then another, and another. . . . Collectively, these blocks become a

---

[231] Other cryptocurrency teams may choose to change their cryptocurrency architectures in the future. So, they could function differently by the time you read this.

chain of blocks called a *blockchain*. One of the most important reasons that blockchains are so secure and immune to hackers, corrupt corporations and governments is because blocks that come later in the chain make it nearly impossible to change blocks that came before them. This concept is easy to understand with the following illustration.

This is why bad guys can't cheat the Gini Blockchain.

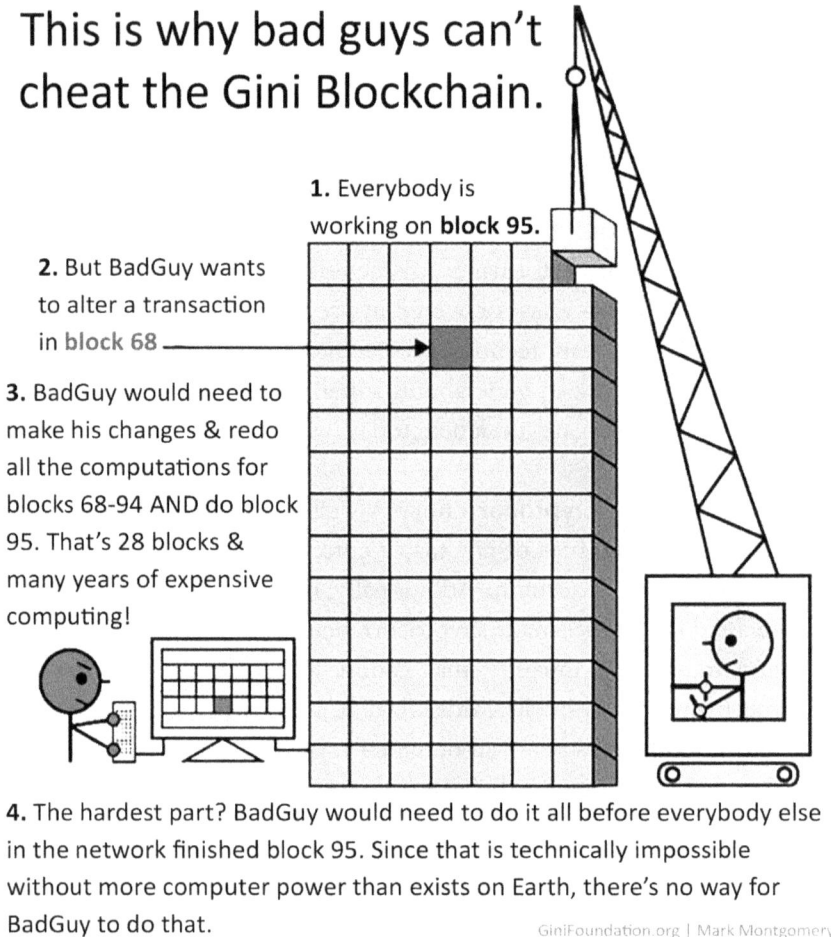

1. Everybody is working on **block 95.**

2. But BadGuy wants to alter a transaction in **block 68**

3. BadGuy would need to make his changes & redo all the computations for blocks 68-94 AND do block 95. That's 28 blocks & many years of expensive computing!

4. The hardest part? BadGuy would need to do it all before everybody else in the network finished block 95. Since that is technically impossible without more computer power than exists on Earth, there's no way for BadGuy to do that.

GiniFoundation.org | Mark Montgomery

Tamper-Proof & Secure: Cryptocurrencies Can Prevent USD Trillions in Fraud. Unlike all other forms of payment that can be manipulated by criminals, monopolistic banks, self-serving corporate executives, and corrupt politicians, cryptocurrencies are stored on a blockchain, which prevents any person or entity from manipulating a transaction record after it has been recorded on the blockchain. Based on the principle that "a penny saved is a penny earned," and considering that the combined cost of fraud in the ecommerce sector, retail sector, mortgage lending industry, corporate accounting, and securities trading is measured in USD trillions, this tamper-proof feature alone is sufficient to substantiate the intrinsic value of a well-designed cryptocurrency.

Transaction Privacy. In a previous chapter, we learned that privacy is one of the most important human rights; and when governments and corporations spy on us, that prevents humanity from holding governments and corporations accountable for their actions. Mainstream cryptocurrencies like Bitcoin and Ethereum do not provide sufficient privacy because they publicly broadcast the account (aka, "wallet") addresses of all transacting parties. In the early days of cryptocurrency development, publicly broadcasting all transaction details was perceived as a strength to ensure transaction integrity and accountability. But as humanity begins to take cryptocurrencies seriously for everyday commerce, privacy becomes a much more important factor; thus, a new approach is needed.

Governments Are Already Creating Crypto Blacklists Today. The U.S. Office of Foreign Asset Control (OFAC) has announced that they will be blacklisting known Bitcoin account/wallet addresses, which destroys the fungibility of those bitcoin.[232,233] The FBI has been secretly doing this for a while already, but expanding the practice to the OFAC means the entire U.S. Government and many other entities will monitor the same global blacklist database. Over time, this will essentially destroy Bitcoin's value and the value of other cryptocurrencies *that publicly broadcast all their transactions* because nobody will want to accept blacklisted coins. This will inevitably amplify all the privacy problems discussed previously, which is why the Gini architecture is designed differently.

Trillions of Digital Spies Are Coming Soon. After governments inevitably convert all their fiat currencies into digital currencies, and

---

232 OFAC FAQs: Sanctions Compliance. (n.d.). Treasury.gov/resource-center/faqs/Sanctions/Pages/faq_compliance.aspx#vc_faqs
233 OFAC's Bitcoin Blacklist Could Change Crypto. (2018, March 24). Coindesk.com/goodbye-fungibility-ofacs-bitcoin-blacklist-remake-crypto

when they use the OFAC database to track previous generation cryptocurrencies like Bitcoin, they will be able to visually track every digital currency unit throughout the world on a digital map *in real-time*. That means there will soon be literally trillions of tiny digital spies tracking every human on Earth in real-time if they're using trackable digital currencies. Every cup of coffee, every drink at the bar, every tank of gas, every pack of condoms, every over-the-counter drug, every financial gift to friends and family, literally *every purchase and transaction* you make will be tracked and attached to your *citizen profile*. Then, anybody with government access or powerful political connections will be able to *subscribe* to your citizen profile and cross-reference your private data with all other government and corporate databases.

Do You Have Blacklisted Coins? There is a 90% probability that *any one* of the USD bills that you have possessed at some point in your life were used in a crime.[234] That means there is a nearly 100% probability that you've possessed USD bills that were used in a crime. But you never needed to care because a dime or a dollar bill do not instantly lose their value simply because they were used by a criminal in the past. In fact, we all expect dollars, dimes and all other physical currencies to be worth the same today as they were yesterday before we received them (excluding inflation). Even if a particular bill's serial number is recorded from use in a crime, there is no economy-wide tracking system (yet) that can detect those bills circulating *in real-time*.

Blacklisted Digital Coins Will Instantly Lose Their Value. Digital currencies tracked through a public blockchain become *tainted* when a government places them on a blacklist. In fact, non-private digital coins do not even need to be used in a crime; the *mere opinion* of a politician or bureaucrat who *claims* the prior coin owner committed a crime is enough to blacklist those coins. If you happen to be the current owner of those coins, their value is instantly destroyed if they're added to a blacklist, regardless of whether you're a criminal or not.

Non-Private Cryptocurrencies Are a Dictator's Dream. Digital currency blacklists enable rogue politicians and bureaucrats to instantly

---

234 90 percent of U.S. bills carry traces of cocaine - CNN.com. (2009). CNN.com/2009/HEALTH/08/14/cocaine.traces.money

turn off your digital money, which is like freezing your bank account, but much more efficient. This is a dictator's dream because it enables them to instantly freeze or completely destroy the money of any person or group that questions their policies or protests against their abuses of political and economic power.

Do We Really Want Trillions of Tiny Spies Embedded in Our Lives? A CNN report revealed that "90% of U.S. bills carry traces of cocaine," yet there is no rational reason to track the owner of every unit of USD currency throughout the entire USD money supply merely because they were used in illegal activities in the past. The same is true for cryptocurrencies. A currency is a unit of account, medium of exchange, and store of value for use in transactions between *private parties*. Transforming every currency unit into a surveillance beacon to create trillions of tiny spies working on behalf of the ever-expanding police state is a guaranteed recipe for even more government tyranny.

Destroying Privacy Is Unacceptable & Unnecessary. With an understanding of the history of government and corporate tyranny, it should be clear that broadcasting all transactions to the public is foolish and naive from a human rights perspective. In fact, losing our transaction privacy enables any politician or corporation to link one of your known transactions to all your other transactions. Clearly, this is an unacceptable trade-off from a human rights perspective. More importantly, it's an *unnecessary* trade-off with Gini's unique technical architecture, which we discuss in a later chapter.

Are Cryptocurrencies a Legitimate Tool to Protect Privacy? Cryptocurrencies *are without a doubt* a legitimate tool to protect privacy because any tool designed to protect human rights that does not infringe on anybody else's human rights is necessary *and* legitimate *by definition*. Regardless of what self-serving banks and politicians may claim, there is absolutely nothing fundamentally wrong or immoral about using any tool to protect our human right to privacy. No terrorist or money laundering threat (which are largely created by misguided politicians) legally or morally supersedes any of our constitutional or human rights. Any attempt to argue otherwise is a defense of tyranny, a sales pitch for self-serving security companies, and a direct violation of everything that America's Founders intended when they fought the American

Revolution and gave us the U.S. Constitution and Bill of Rights.

Humanity's Response to Tyranny

Should Humanity Submit to Tyranny or Resist Tyranny? When faced with the sobering realities of our world today, humanity has two basic options: Submit to tyranny or resist tyranny. But how do we "resist tyranny"? It's not enough to buy some bitcoins, scurry into the shadows like rats, and hope for the best—major banks and governments are working together to shut down all gateways into the crypto ecosystem, which will be easy for them to do with mainstream cryptocurrencies like Bitcoin and Ethereum due to their privacy-*unfriendly* architectures.

Words Are Not Enough. It's not enough for libertarians to scream, "Government, leave me alone!" because the degradation of humanity by A.I. and corporate cannibalism cannot be resolved by the absence of government alone. To the contrary, the survival and stability of human societies depends on the following fundamental truth.

Active Resistance *and* Crisis Intervention is the Only Viable Path Forward. In the age of A.I. and transnational cannibals, Libertarianism is not enough. Hope is not enough. The only realistic path to meaningful human liberty and economic survival in the 21st Century and beyond is to actively resist tyranny by injecting specific humanitarian and ecosystem stability mechanisms into our economic, geopolitical, and technological systems. This is the only way to achieve meaningful ecosystem stability, balance, and sustainability. Only with an ecosystem-wide perspective will humanity survive the rapidly expanding economic and political onslaught of A.I.-powered systems, the transnational cannibals that control them, and the politicians who increasingly collude with corporations and A.I.-powered systems against the best interests of humanity.

The Future is Based on the Choices We Make Today. Over the past several generations, human civilization has been brainwashed with soul-less consumerism, exploitative corporate greed, and manipulative government propaganda, which causes many people to believe that justice, equity, and fairness are no longer possible. This is not true. Change can happen quickly. After a tipping point occurred, the USSR rapidly collapsed in just a few weeks after generations of tyranny. The economic and political systems that exist today are the result of specific choices that we, as a species and as human societies, have made. We can make different choices. Gini represents a different choice and a different path—a path toward meaningful justice, equity, and fairness.

Key Points

- **Cryptocurrencies Have Tremendous Value.** The intrinsic value of a well-designed cryptocurrency is different but complimentary to the value of precious metals. However, unlike precious metals, a privacy-assured cryptocurrency is virtually impossible for misguided politicians to confiscate or destroy.

- **Any Tool Used to Protect Human Rights is Legitimate *by Definition*.** Never let anybody tell you that cryptocurrencies are only used by criminals and scammers. Transaction privacy is a human right and an essential ingredient of every free and democratic society. What we buy with our own money reveals our most private thoughts, actions and intentions. Without transaction privacy, we have no way to protect ourselves from the spying, manipulation, and sabotage of politicians and corporations who will inevitably use their wealth and power to prevent citizens from protesting their oppressive policies or holding them accountable for their tyrannical actions.

- **Cryptocurrencies Are Unstoppable.** Banks, governments, and special interest groups have done everything *politically possible* to stop the spread of cryptocurrencies. However, broken tax systems will give politicians new incentives to create illegitimate laws and use physical violence in the future. Additionally, the banks are now trying to co-opt cryptocurrencies and steer the general public toward their own bastardized blockchains. They recognize that cryptocurrencies are unstoppable and the only thing they can do is try to inject their toxic DNA into the crypto ecosystem. All liberty-loving humans should resist their self-serving manipulation.

- Chapter 7 -
Crypto Monetary Policy

"The U.S. government has a technology, called a printing press . . .
that allows it to produce as many U.S. dollars as it wishes
at essentially no cost."
— Ben Bernanke, former Federal Reserve Chairman

A crypto *or* fiat currency without an effective monetary policy is like a ship without a compass. The ship might *accidentally* reach its destination if all the trade winds happen to be blowing perfectly in the right direction, but there's a much higher probability that it will become shipwrecked on the rocks of the currency graveyard. Every major cryptocurrency project today is plagued by poorly designed monetary systems, which prevents them from ever becoming truly equitable and viable currencies for global commerce. This chapter explains why this is true and why Gini is different.

Cryptoeconomic Ideologies

Ideology vs. Reality. Many humans confuse ideology with reality, which often spawns a tribe mentality. In this context, the tribe mentally occurs because many people get confused and frustrated when they assume economic and political systems are separate domains and/or they get overwhelmed by all the noise and propaganda from special interest groups, banks, corporations and politicians. When humans are confused and overwhelmed, their emotions get triggered, which creates fears and insecurities. This creates an impulse to seek safety with a familiar tribe (e.g., Keynesians, Austrians, Marxists, Democrats, Republicans, Liberals, Conservatives, etc.). But the safety of their tribe is just a regression back into a comfortable existential bubble, which results in a missed opportunity to expand their awareness of the broader universe outside that bubble.

All Economic Ideologies Are on the Same Continuum of Political Economy. We see the tribe mentality emerge when people argue over capitalism vs. communism, democracy vs. authoritarianism, free markets vs. coordinated markets, purely free trade vs. import substitution industrialization (ISI), supply- vs. demand-driven economies, etc. When people engage in these endless debates, it means they don't understand that all these concepts are merely different points along the same continuum of Political Economy. Where a society, community, or country exists along the Political Economy continuum at any point in time depends on their phase of economic development and how their citizens prioritize ecosystem sustainability vs. unlimited power and profit for a few.

Maximizing Production _is Not Equal_ to Maximizing Human Welfare. A huge false assumption that virtually all mainstream economists have made since the late 1970s is that GDP is all that really matters. They claim that as long as a country's GDP is growing, then all the wealth will magically _trickle down_ to the masses and there will be _more winners than losers_. Similarly, many people in cryptocurrency communities today still believe that all the crypto whale wealth will trickle down to the crypto masses when the whales start selling their cryptocurrencies.[235]

Trickle-Down Economics in the Real World. A decade after Bitcoin was born, the trickle-down cheerleaders still believe, as a matter of _faith_, that the crypto markets will eventually become less concentrated over time. If they understood economic history and the data associated with the decline of real median income and wealth in the U.S. and other advanced economies, they would not be so easily deceived by this fantasy. In other words, there is no credible evidence that trickle-down economics _has ever worked_ in the fiat world; so, there is no rational reason to believe it will ever reduce the concentration of wealth and power to a sustainable level in the crypto world.[236]

The following charts illustrate this truth.

---

235 "Whales" are typically early investors in a cryptocurrency project, which exert overwhelming control over every cryptocurrency ecosystem today.
236 In addition to these charts, see the article "Collapse of the Human Labor Force" at Eanfar.org to see how and why the _official_ government unemployment rate statistics are grossly misleading.

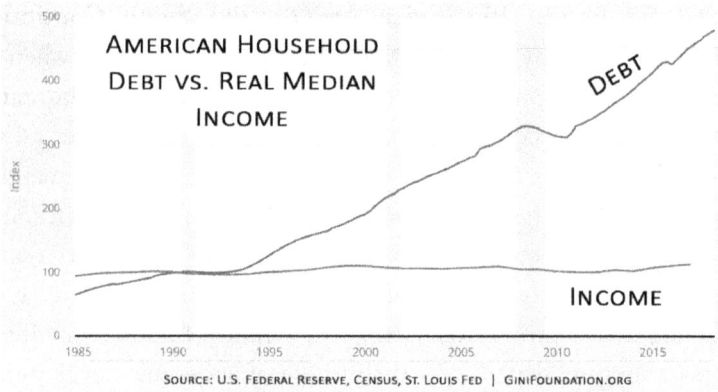

AMERICAN HOUSEHOLD DEBT VS. REAL MEDIAN INCOME

DEBT

INCOME

SOURCE: U.S. FEDERAL RESERVE, CENSUS, ST. LOUIS FED | GINIFOUNDATION.ORG

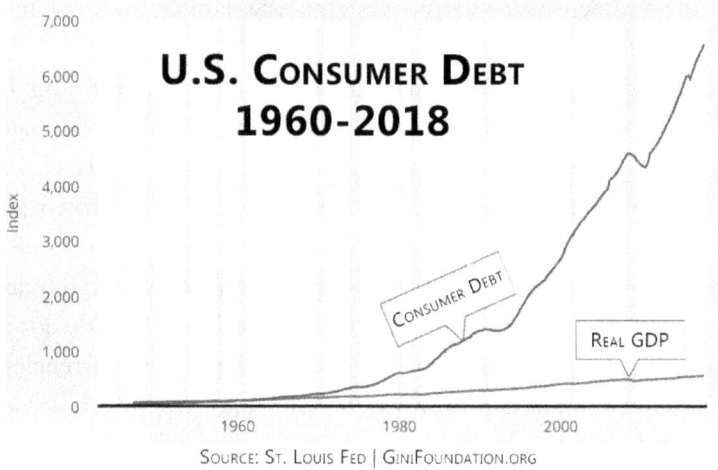

U.S. CONSUMER DEBT
1960-2018

CONSUMER DEBT

REAL GDP

SOURCE: ST. LOUIS FED | GINIFOUNDATION.ORG

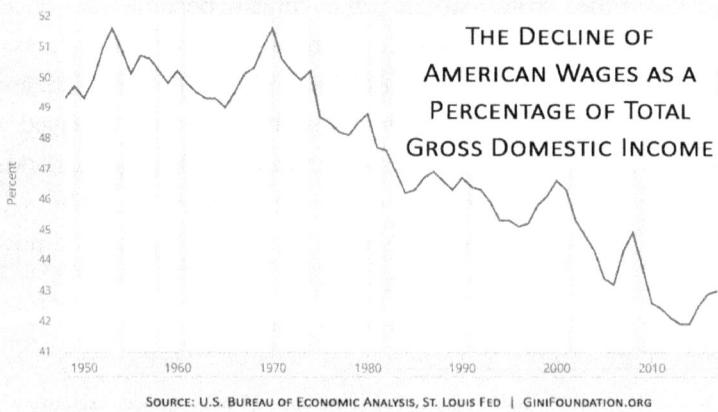

THE DECLINE OF AMERICAN WAGES AS A PERCENTAGE OF TOTAL GROSS DOMESTIC INCOME

SOURCE: U.S. BUREAU OF ECONOMIC ANALYSIS, ST. LOUIS FED | GINIFOUNDATION.ORG

AMERICAN LABOR FORCE PARTICIPATION RATE VS. POPULATION GROWTH

POPULATION

LABOR FORCE

SOURCE: U.S. BUREAU OF LABOR STATISTICS, U.S. BUREAU OF THE CENSUS, ST. LOUIS FED | GINIFOUNDATION.ORG

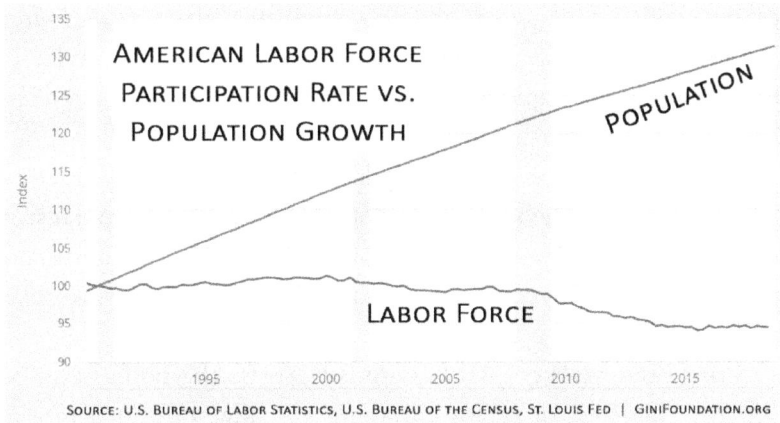

Unprecedented Debt Masks a Disturbing Reality. The only reason that Americans and many people in the Western world have been able to experience a relatively high quality of life over the past generation is because they have been drowning in debt. That debt has masked a disturbing reality: Gigantic banks and corporations and a tiny number of their largest shareholders have siphoned off the vast majority of *real* wealth from humanity. In fact, without accumulating more debt than any civilization in human history, between 80-90% of all Americans today would be living in developing world conditions. The same fate is inevitable for humans in many other countries that have blindly followed American-style corporatocracy.

Stock Market Gains Are a Fantasy for Most People. Many people assume that the temporary gains in the stock market actually help the middle class. That fantasy is easily debunked with the real, long-term economic data, as illustrated in the charts above. For all these reasons and many more, the Gini cryptocurrency technology is based on much more sustainable and equitable monetary principles.

Stagnant Middle-Class Incomes Prove Trickle-Down Economics Doesn't Work. Aggregate economic metrics like GDP (or crypto market cap) reveal nothing about how wealth and power is distributed throughout a population; and *the distribution of wealth and power in every population is what determines the systemic health of every economy, including a crypto economy.* This is why measuring the Gini Index and *median income* (*not* per-capita income) is so important. Measuring those factors, we can

147

see that high Gini Indexes and the completely flat and declining real income of the middle classes in nearly all rich countries today (while their GDP's have exploded) proves that trickle-down economics is a fantasy. Yet, all mainstream fiat and crypto economic theories, models, and cryptocurrency projects are built on this fantasy today. Predictably, these are the same theories, models, and fantasies that have transformed our global economy into the debt-choked, unstable, unsustainable, oligarchic system of corporate greed and economic injustice that it is today.

Pure Free Markets Are a Fantasy. Anybody who has ever been a lobbyist, politician, or senior corporate executive operating in a global market knows that pure free markets are a fantasy. This fantasy completely ignores real-world geopolitics and fallible human nature. It ignores the fact that there will *always* be *government* intervention (both in fiat economies and crypto economies) because all human-controlled systems tend to concentrate unsustainable wealth and power toward the class with the most political power (whales, oligarchs, politburo commissars, despots, etc.). The only way to prevent this is to engineer counter-balancing mechanisms *directly into the system* to prevent the concentration of wealth and political power from distorting the system in the first place.

If Government Intervention is Inevitable, How Should Governments Intervene? Given that politicians *will always intervene* to skew the distribution of a society's wealth toward the most politically connected creatures, the primary *real-world* question that we should ask is this: Will governments intervene to amplify the economies of scale of the *whales* (i.e., transnational cannibals in the fiat world and crypto whales in the crypto world) or will governments intervene to create *ecosystem balance* so that small- to medium-sized companies and the middle class can achieve a sustainable level of wealth and political influence within the ecosystem?

How Should Economic Value be Distributed? Whenever a society asks a question about what *should* be, the question obviously implies certain trade-offs and value judgments. Logically, these judgments should be based on a society's *highest priorities and purpose*. When rules and institutions are not aligned with a population's highest priorities and purpose as expressed by a majority vote of that population, then that population will be constantly plagued with conflict, which will constantly erode—and ultimately destroy—a large portion of the value flowing through that population.

Inequitable Economies Are Unsustainable Economies. To answer the question—"How should economic value be distributed?"—we must acknowledge a fundamental principle in human nature and Political Economy: Large human populations will not tolerate or embrace any socioeconomic system that does not allow them to create and distribute value *equitably*. Thus, value must be distributed *equitably from the perspective of a majority of people in a population;* otherwise, the system is socially, politically and economically unsustainable.

What Does "Equity" Mean? The concept of "equity" has been an integral feature of the U.S. legal system from its founding. American courts of equity are based on the centuries-old English Court of Chancery, which today is a division of the British High Court of Justice. The dictionary definition of "equity" encompasses the principles of fairness, impartiality, justice, proportionality, and egalitarianism. Most civil matters associated with corporations, family law, and bankruptcies in the United States are adjudicated in *courts of equity*, which are distinct from *courts of law*. In courts of equity, the judge is required to exercise significant discretion over each case because the amalgamation of legal contracts, financial and material assets, and a wide spectrum of human activities often collectively result in a highly complex situation that cannot be reduced and codified into concise laws.

What Does "Equality" Mean? Although these words are often used interchangeably and confused with one another, the concept of "equality" is very different from the concept of "equity." The dictionary definition of "equality" encompasses concepts like "quantity" and "uniformity," which are more precisely understood as mathematical principles rather than philosophical principles. A mathematical equation like $9 + 1 = 10$ is true because the quantities on both sides of the equal sign are perfectly equal by definition. No matter how complex the equation, both sides must be equal, otherwise the equation is logically false. Thus, the equation also has uniformity as the quantity of both sides must be perfectly identical for it to represent something that is logically true. This is the essence of the principle of *equality*. So, we can see that *equality* is fundamentally about counting and measuring things to achieve perfect uniformity.

Proportionality is the Bridge Between Equality and Equity. Virtually everybody can embrace the concept of equality *in essence* without expecting equality *in quantity*. For example, all humans are equally human even though they don't all have equal body weight. All racial and ethnic groups can be given equal voting rights without expecting all groups to cast an equal number of votes in an election. Men and women can enjoy equality under the law without expecting them to have an equal number of chromosomes. The equal essence of things without having an equal quantity of things exists all around us without diminishing our sense of justice. This is because the human mind unconsciously equates justice with *proportionality, not equality*.

Policies & Systems that Emphasize Proportionality Create Equity and Unite Humanity. Understanding the distinction between equity, equality and proportionality is the key to building sustainable economies, fair and just societies, and uniting humans across the ideological and socioeconomic spectra. Policies that invoke a spirit of equality to protect the essence of humanity (life, liberty, pursuit of happiness, etc.) while creating a strong systemic bias in favor of small- to medium-sized stakeholders is how governments and cryptocurrency projects can achieve balance between equality and proportionality within an economy and throughout every capitalistic society. This equality-proportionality balance is what *creates equity* in all human relationships and transactions, which is the most sustainable way to build blockchain platforms, human economies and societies.

Unbridled Capitalism *is Not Equal* to Human Liberty. Many Ayn Rand lovers (including me when I was younger) believe that true freedom comes from pure free markets. After many years of traveling and studying different economies, talking with political leaders in different countries, operating companies in different economic systems, dealing with corrupt tax and regulatory regimes that are engineered to benefit the largest transnational corporations . . . I realized that *unbridled capitalism is not the same as human liberty*. In fact, contrary to Ayn Rand's and Hayek's extremely narrow view of Political Economy, unbridled capitalism quickly destroys human liberty because every economy is inevitably touched by government intervention. This inevitably creates a systemic bias toward the most politically well-connected business groups in a society unless strong counter-balancing mechanisms are *engineered into the system.*

The Consequences of Poorly Designed Monetary Policies. Without conscious technical design choices to achieve ecosystem stability, the dynamics above inevitably cause the wealth of fiat and crypto economies to flow to the most politically connected segment of the population based on inherited aristocracy, luck of birth time and place, private voucher sale invitations, and many factors that have nothing to do with merit or the actual value they contribute to society. This is exactly what we already see in all major fiat and cryptocurrency economies today, which is why a small number of anonymous whales dominate all cryptocurrency markets today. This is not an inevitable feature of human existence, *but it is an inevitable consequence of poorly designed monetary systems.*

Investor Capital is Only One Form of Value. Investor capital is only one of many forms of value that a society should economically recognize and reward. Just like with land value, the value flowing through every economy is created by many types of people with many types of skills coming together around land, capital, and labor to create value that did not exist previously. The value of capital is already well-compensated in our world today, but labor value is being destroyed due to the rapidly expanding power of A.I., robots, short-sighted economic policies, and the coercive power of owners of land and capital. These accelerating trends are disturbing because labor value is an integral part of the Capital-Labor Duality, which is the foundation of every capitalistic economy.[237] Without distributing value to labor equitably, an economy will inevitably collapse. This will still be true in the future when A.I. becomes more ubiquitous.

Most Cryptocurrencies Today Are Built on a Broken Fiat Economic Ideology. All substantive cryptocurrency projects today are

---

237 The Capital-Labor Duality is discussed on the Gini website (GiniFoundation.org/kb/capital-labor-duality) and more thoroughly in my previous book, *Broken Capitalism: This Is How We Fix It.*

being built on the same broken economic ideology that is destroying the global fiat economy today. They are simply trying to create their own crypto empires to replace the fiat empires. This means the long-term outcome of their cryptocurrency science projects will be the same: The flow of value (wealth, income, opportunities, etc.) in their ecosystems will mirror the concentration of power that dominates all fiat and crypto economies today. Trickle-down economics will not fix their problems, which means their problems will never go away.

Economic Freedom is Not Equal to Business Freedom. Confusing business freedom with economic freedom inevitably *destroys the economic and political freedom of the general population* while it protects the *business freedom* of a small number of oligarchs. This is what happened in Italy during Mussolini's dictatorship (1925–1943), which Hitler then emulated in Nazi Germany (1933–1945). See the following extremely well-documented books (among others) in the Gini Book List, which substantiate this truth.[238]

- *Business as a System of Power*
- *Wall Street & the Rise of Hitler*
- *Tragedy and Hope: A History of the World in Our Time*

Proto-Corporatocracy. Before the economic system called "corporatocracy" fully emerges in a country, the largest shareholders and executives in the largest corporations use their wealth and power generated from their *business freedom* to redirect the distribution of wealth throughout the entire society back to themselves. This gives them economic power to purchase biased politicians, laws, and regulations that enable them to perpetuate their business and economic wealth and political power over generations. This is how capitalism mutates into *corporatocracy* and how gigantic corporations then provide the financial resources for charismatic tyrants like Hitler to ascend to positions of dictatorial political power.

Corporatocracy is the Economic Engine of Fascism. Contrary to Ayn Rand's fictional books that have very little basis in economic

---

238 The Gini Book List can be found here: GiniFoundation.org/book-list

reality, giving whales (transnational cannibals or crypto whales) the freedom to trample all over an economy is not "economic freedom." In fact, this is technically *corporatocracy*—the economic engine of fascism—which protects the *business freedom* of a tiny number of predatory oligarchs while destroying the *economic and political freedom* of the general population.

The Sad History of Wealth & Power Concentration. Most of the problems described in this book are obvious to anybody who has ever studied the economic history of the Gilded Age in the United States; the concentration of wealth and power in Nazi Germany and pre-WWII Italy; pre-Soviet tsarist Russia *and* post-soviet oligarchic Russia today; the totalitarian power in the Gulf Cooperation Council (GCC) states since they discovered oil and natural gas; the long and sad history of economic oppression in virtually every Latin American and African country; the rapid concentration of wealth and power in China, which is more capitalistic in many ways than the U.S. today; and the obscene concentration of wealth and power in the United States today, which is destroying the U.S. economy and political system. All these dynamics are the inevitable consequence of *American-style capitalism*, which is really just imperialistic mercantilism in disguise, which Adam Smith despised.

Do we really want to perpetuate this cycle of mercantilist wealth and power concentration when we know it will inevitably end in violent revolutions and possibly World War III?

Humanity vs. the Whales

Different blockchain projects have different goals and must deal with different legacy issues. For example, Satoshi Nakamoto had to own virtually all the circulating Bitcoin money supply in the beginning because only a tiny number of people truly understood and appreciated what he was building. Unfortunately, today, approximately seven Bitcoin mining pools dominate the entire Bitcoin ecosystem both economically and politically because of the structural incentives created by Bitcoin's architecture.[239] This historical baggage prevents Bitcoin from being used

---

239 This is discussed in more technical detail in the next chapter.

as a viable currency for global commerce because the Bitcoin *whales* often manipulate the Bitcoin market with wash trades, bear raids, pump-and-dump tactics, and other market manipulation, which creates high volatility. Additionally, the whales' political dominance over Bitcoin's ecosystem development prevents Bitcoin from evolving in ways that would benefit humanity more than the whales.

Trapped by the Whales. Nakamoto's decisions are understandable in their historical context, but since the launch of Bitcoin, virtually all other major blockchain projects have made irreversible monetary policy mistakes. These mistakes lead to predictable results: High volatility, high concentration of cryptocurrency wealth and power, reduced trust in the blockchain team, and less flexibility to evolve in ways that would benefit humanity. Thus, they are trapped within the expectations and self-interest of the whales. Specifically, when a blockchain team allows their capital structure (i.e., the investments received by early investors) to distort their currency distribution and long-term wealth distribution, the blockchain team is forced to feed the whales at the expense of the minnows (the general public). This produces socioeconomic outcomes that are even worse than the egregious imbalances of wealth and power that are destroying capitalism in fiat economies today.

Whales and Their Sell-Walls. To visualize how the whale problem creates volatility and distorted market dynamics, the following chart illustrates a huge *sell-wall*, which was created by one of the whales in the Cardano/ADA market on March 1st, 2018.[240] This problem exists in all major cryptocurrency markets today and it occurs frequently whenever the markets start trending lower. As prices gradually decline from broader supply-and-demand forces, the whales get spooked; then they're naturally motivated to preserve their massive amounts of wealth by liquidating their positions, regardless of the impact on the broader ecosystem.

---

240 See the Gini Decentralized Exchange page on the Gini website to see a video of "Spoofy," which is an automated A.I. bot manipulating the Bitcoin market in real-time.

A Sell-Wall

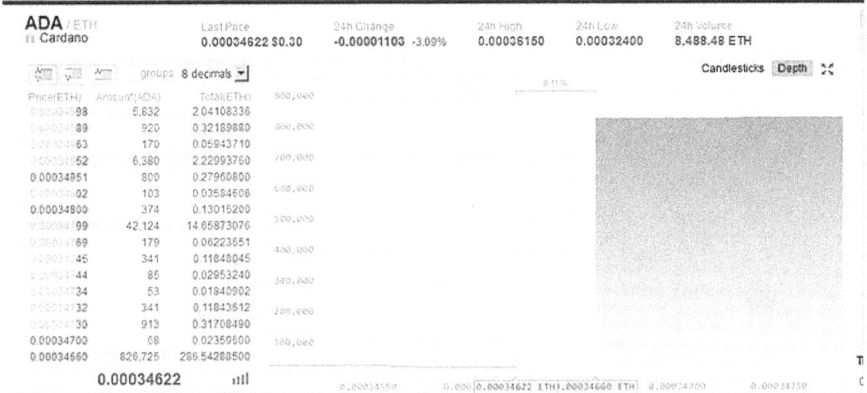

ADA /ETH :: Cardano	Last Price 0.00034622 $0.30	24h Change -0.00001103 -3.09%	24h High 0.00036150	24h Low 0.00032400	24h Volume 8,488.48 ETH

Price(ETH)	Amount(ADA)	Total(ETH)
0.00034998	5,832	2.04108336
0.00034989	920	0.32189880
0.00034963	170	0.05943710
0.00034952	6,380	2.22993760
0.00034951	800	0.27960800
0.00034902	103	0.03594606
0.00034800	374	0.13015200
0.00034999	42,124	14.65873076
0.00034769	179	0.06223651
0.00034745	341	0.11848045
0.00034744	85	0.02953240
0.00034734	53	0.01840902
0.00034732	341	0.11843612
0.00034730	913	0.31708490
0.00034700	68	0.02359600
0.00034660	826,725	286.54288500

0.00034622

A single 826,725 ADA whale trade represents over 3% of the entire 24-hour trading volume. (Chart credit: Pesuazo in the Cardano Forum)

What is a "Sell-Wall"? The big rectangle in the chart above is called a "sell-wall" because it represents a veritable mountain of price-suppressing liquidity that will be dumped into the market if a certain price is reached. In this case, the whale's sell order will be executed if the price of the ADA cryptocurrency reaches 0.00034660 ETH. That single trade represents *over 3% of the entire 24-hour trading volume* in that market. Considering that there are thousands or millions of trades per day in the most popular markets, 3% for a single trade *is a massive portion* of the daily trading volume. The mere existence of such a massive sell order impacts the price of the currency; and when sell orders of that size are executed, they take the entire market down to much lower price levels.[241]

Bear Raid & Pump-and-Dump Attacks. When whales have the power to deliberately or inadvertently distort markets, unsuspecting stakeholders have no idea why the value of their wealth suddenly collapses.[242] During these market collapses, whales profit from their sell orders; *then* they can gobble up even more currency after they cause the market price to plunge. *Then*, they create a *buy-wall*, which pushes the

---

241 Here's a good explanation of sell-walls: Nugget's News Australia. (2017). Buy/Sell Walls . . . What You Need To Know. Youtube.com/watch?v=LSkEWrCT-eo&feature=youtu.be&t=419
242 A dramatic example of this common market manipulation technique can be observed here: "Watch Spoofy place $40,000,000 USD in fake orders." Youtube.com/watch?v=C-JBGK4BuME

price back up rapidly, giving them an opportunity to pump the price up until they're ready to dump another huge load of crypto into the market. In securities trading, this illegal strategy is called a "Bear Market Raid" on the way down and a "Pump-and-Dump" on the way up. Regardless of their legality, these and other illicit strategies are only possible when a market is dominated by a relatively small number of whales.

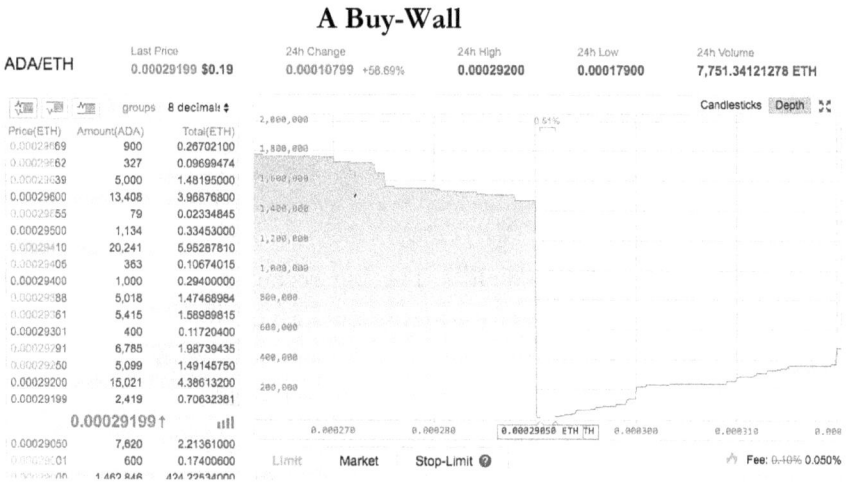

A Buy-Wall

ADA/ETH	Last Price 0.00029199 $0.19	24h Change 0.00010799 +58.69%	24h High 0.00029200	24h Low 0.00017900	24h Volume 7,751.34121278 ETH

Price(ETH)	Amount(ADA)	Total(ETH)
0.00029869	900	0.26702100
0.00029862	327	0.09699474
0.00029839	5,000	1.48195000
0.00029600	13,408	3.96876800
0.00029655	79	0.02334845
0.00029500	1,134	0.33453000
0.00029410	20,241	5.95287810
0.00029405	363	0.10674015
0.00029400	1,000	0.29400000
0.00029388	5,018	1.47468984
0.00029361	5,415	1.58989815
0.00029301	400	0.11720400
0.00029291	6,785	1.98739435
0.00029250	5,099	1.49145750
0.00029200	15,021	4.38613200
0.00029199	2,419	0.70632381
0.00029199↑		
0.00029050	7,820	2.21361000
0.00029001	600	0.17400600

A single 1.5 million ADA whale trade represents 5.5% of the entire 24-hour trading volume.
(Chart credit: Skryb on Reddit)

Bad Monetary Policies Create Human & Systemic Corruption. No matter how exciting a blockchain technology might seem, if it's dominated and controlled by a small group of whales, it will always be highly vulnerable to human and systemic manipulation and corruption. Additionally, regardless of their intentions, if a blockchain team or their investors come across as greedy, they will not gain the trust of the general public. Even when people don't fully understand the technology, they're still usually smart enough to see through empty platitudes and shallow mission statements that are not truly aligned with a blockchain project's currency distribution and monetary policy.

Alignment of Interests. Regardless of their intentions, virtually all other cryptocurrency project teams have chosen to make monetary policy decisions based on factors that have been substantially dominated by profit-seeking whales and insiders. This undermines the spirit, intent, and systemic integrity of everything they're doing. Time and again, we see excuses similar to this: "Our investors needed a return on their investment; so, this is the crypto distribution we are forced to work with." Whether this is true or not is impossible to know because the interests of their investors are aligned with the interests of their founders, not the general public. So, it's convenient for them to blame the investors and capitalism in general.

Blockchain Teams Should be Public Stewards, Not Profiteers. We know from real-world experience that the "investors made us do it" excuse is usually just another way of saying, "We don't know how to raise money any other way; so, let's take the money and hope for the best." If they were building a for-profit corporation, we would have no problem with their *investors-are-king* logic, but they usually claim to be building world-changing economic systems for humanity, not for a small group of insiders; so, within that context, it is not acceptable to allow a small group of whales to dominate their entire ecosystems.

Earning Public Trust Is the Most Sustainable Way to Earn a Profit. If a blockchain team truly understands how their technology will be beneficial to humanity, then they should understand how to articulate to investors the importance of their role as public stewards of that technology, which should be perceived *primarily* as a public utility or trust, and *secondarily* as a profit-generating investment. To be clear, we're not opposed to investors receiving a strong return on their investment, but a public blockchain's priorities should be *to the public first*, not their investors. By serving as effective stewards on behalf of the general public, investors will receive a handsome return on their investment *as a natural byproduct of being effective public stewards of their projects.*

Purpose-Driven Monetary Policy

To put everything described in this chapter into perspective, below is a summary of an epic months-long debate that I had with several members of another major cryptocurrency community. This was while I was heavily invested in and deeply passionate about that project. My sadness about the project's broken monetary policy was the source of my comments during this debate.[243]

========== BEGIN EXCERPT ==========

Our Project's Purpose. Somebody in our community recently said that early investors in this project deserve to control the ecosystem because they invested first and took early risks. That would be true if we were talking about a *for-profit* corporation like Apple or Goldman Sachs, but that's not how this project has been presented to the world. **This project is not intended to be a for-profit corporation; it's intended to be a trust-less socioeconomic platform and crypto-economy for humanity, upon which many smart-contract-based socioeconomic tools and institutions can be built.**[244] *That's* why I've invested in this project; profit considerations are purely secondary to *that primary purpose.*

Lucky Birth & Geography Should Not Create a Crypto Aristocracy. Every violent revolution in human history has been caused by oppressed people rising up against the tyranny of inherited, *concentrated* aristocratic wealth and power. That tyranny is *always* rooted in poorly designed economic systems and incentive structures. If we assume that it's OK to accept a despotic level of economic/political control over this ecosystem (*over the long-run*) simply because the early investors were born in the right place at the right time, then we are ignorant of history and ignorant of the *inevitable* catastrophic consequences of oligarchic/aristocratic domination of an economy.

Prioritizing Purpose Over Profit *Does Not* Eliminate Profit. I am a capitalist, an investor, a corporate executive in the FinTech industry, a published book author in the field of International Political Economy, among other things that give me "capitalist street cred," but I also understand how to *distinguish between profit and purpose*, just like Adam Smith did. So, of course I understand that investors in any project should receive a positive return on their investment, *but that's not the primary purpose of a public cryptocurrency*; and we can be certain the early

---

243 I have edited these excerpts only to remove personal and organizational names to save them from unnecessary attacks. I still wish that team great success in all their endeavors.

244 After a while I realized that this was my interpretation of the project's purpose based on many of the comments made by the founders, but the reality turned out to be substantially different.

investors know that. But let's define "profit" here to make sure everybody can see just how "communist" we are for *daring* to prioritize purpose over profit. . . .

How Much Profit Have the Whales *Already* Received? [At this point in the debate, I presented all the calculations to substantiate the following astronomical ROI figures that the early investors received. Those calculations are omitted here to avoid publicly revealing any particular investors and stakeholders in that project. Now my debate with them continues below. . . .]

- Earliest Investors' ROI: 39,000%
- Avg Profit for All Early Investors: 34,000%

And if we performed these calculations earlier this year at peak prices, the ROIs would be ~3x higher.

Connecting Reality to Purpose. Given the following facts . . .

- The *primary purpose* of this project is to build a trust-less socioeconomic platform for humanity.
- The *primary purpose* (by definition) of every project should be given a *higher priority* than all other factors.
- **The whales have *already* received at least 34,000% profit** (unrealized); and some have received over 100,000%, depending on when and how much crypto they've already sold. That's *already* more profit than most investors ever get in their wildest dreams.

Are We Being "Communist" or Threatening the Sanctity of Capitalism? Is anybody in this community going to claim that *capitalism is being threatened* because many of us in the community want the team to prioritize *the purpose* of this project above profit motives? Is anybody in this community going to claim that investors are being *injured* merely because we want to *gradually* reduce the concentration of wealth/power of the whales over a *gradual* period of time *after* they've already received at least a 34,000% profit (as of this moment), which will most certainly exceed *1,000,000%* profit by the time they actually liquidate their crypto positions according to the rational framework I presented previously?

Do We All Understand Economic History & the Fundamental Purpose of This Project? If anybody in this community still believes that the whale debate is about communism vs. capitalism or thinks that the risks I described previously are unrealistic or over-exaggerated in any way, then they have never actually read Adam Smith's *Wealth of Nations* or his *Theory of Moral Sentiments* (or any other serious economist), they know nothing about economic history, and they have no meaningful understanding of the catastrophic consequences of allowing an oligarchy to own and/or control an entire economy. . . .

Reviewing Our Goals as a Community. In life and in business, whenever we establish priorities about how to spend our scarce time and resources, we must make trade-offs based on the goals we want to achieve. IMHO, the *highest priority* and *primary goal* should be to *earn and preserve* the trust of every human on Earth today who is evaluating this project's *actual* level of wealth / power / network decentralization. Why should that be our highest priority (excluding the obvious technical priorities)? Because every serious student or scholar of Political Economy knows that this crypto economy is doomed *over the long-run* if it's controlled by a relatively small number of whales that can control all the decision-making processes. Anybody who does not grasp these realities is either focused on short-term speculation or they have no meaningful economic and geopolitical education.

========== END EXCERPT ==========

Broken Monetary Policies Destroy Community Enthusiasm. The excerpts above represent only a tiny fraction of the 100s of man-hours that went into that debate. In return, I was attacked by the crypto whales and/or their trolling cronies and the core team ignored the concerns of hundreds of us in that community who were previously deeply passionate about that project. In fact, I was the most "liked" contributor in that community and many of us were deeply involved in several projects based on that technology until we realized how unnecessarily concentrated their crypto money supply actually is.

Purpose-Driven, Economically Sustainable Capitalism. It should be clear based on everything presented in this book so far that the Gini team is not opposed to investors who want a profitable return on their investment. We are not opposed to the core Smithian principles of free-market capitalism. In fact, the entire purpose of my previous book was to prevent capitalism from completely collapsing. However, we are absolutely opposed to launching a *public* cryptocurrency on an oligarchic foundation, controlled by whales seeking perpetually increasing profits of 100,000% to 1,000,000% and beyond, which will inevitably create perpetual volatility, perpetually broken incentive structures, perpetual wealth and power inequities. . . .

In one of our many debates, I said:

This Is Supposed to be a Trust-Less System, but How Can We Trust Anonymous Whales? *I'm sure they're all nice and trustworthy mammals, but one of the most important goals of this project is to achieve a trust-less economic system that does not require faith in fallible human nature or unknown whale intentions. I'm not questioning anybody's intentions, but these anonymous whales currently thwart the spirit and intent of that goal. At this moment, it appears that these whales represent an even greater potential threat to this ecosystem than the Federal Reserve is to USD today because the Fed is at least theoretically accountable to the will of the American people indirectly through the election of the president and members of Congress, who appoint the Fed Chairperson. However, these unelected, anonymous whales are protected and entrenched by anonymity and contract law, which supersedes all other mechanisms of democratic governance.*

In response, one community member said:

This argument seems to contradict itself in the first sentence: "This Project Is Supposed to be a Trust-Less System, but How can we trust anonymous whales?"—That's the point, we don't have to. It's trust-less; whales or no whales. Their "unknown" intentions don't matter unless they all collude.

That person made several other comments, all of which revolved around the idea that the whales have good intentions because nobody would want to hurt the value of their own cryptocurrency holdings. My response below to his comments may help others to understand why his comments demonstrate a deep misunderstanding of human incentives within economic and political systems, and in particular, the problems within all crypto markets today.

========== BEGIN EXCERPT ==========

There are no paradoxes in life. What humans perceive as a paradox is actually just a collision with our own ignorance. In this case, there is no contradiction [in my statement] if you truly understand my point: A system can be *intended* to be trust-less, but a flaw in the system's implementation can thwart its *actual* trustworthiness. More specifically:

1. The technical integrity and trustworthiness of a blockchain is undermined when the 50%+1 attack risk exists (and it does in this case).

2. The integrity and trustworthiness of a *crypto market* and corresponding economy is undermined when the concentration of crypto wealth gives whales enough power to manipulate the market (as proven in the whale manipulation charts).

3. The integrity and trustworthiness of a blockchain's *political governance* (i.e., technical development decisions, project voting, PR strategies, minting rewards algorithms, allocation of scarce team resources, crisis management strategies, regulatory and government relations strategies . . .) is undermined when a non-trivial portion of the community has *reasonable doubts* about the long-term viability of the project and/or the integrity of its crypto market due to the preceding two vulnerabilities.

Thus, on every meaningful level—technical, market, and political—a high concentration of wealth in any economic/market system completely undermines the integrity of that system. In response to these points, there are two general good-faith responses:

1. "We trust the team and whales; so, don't worry, be happy and

enjoy the ride."

OR

2. "We love the team, but the high crypto concentration has many unintended long-term consequences that represent real risks to the viability of our project and community; therefore, we should take these concerns seriously and do everything possible to resolve these problems ASAP."

The people who take your position have chosen the first option. I've lived long enough to know that option just kicks the can down the road. So, I choose the second option. [That community member then said the following.]

Having a lot of money doesn't let you 'control' the economy.

Yes, it certainly does. Have you ever worked on Wall Street? In Canary Wharf in London? Have you ever lived in the U.A.E., Saudi Arabia, Mexico, Russia, DRC, or any country in which the financial system is blatantly dominated *and controlled* by whales? Did you see how less than 10 people who substantially controlled over 70% of all the wealth within the American economy were able to dictate to over 300 million Americans that their taxes would be used to bailout the banks in 2008? How did they get that power and leverage over the entire ecosystem? They controlled the wealth because they controlled the laws that govern the way the system works; that's the nexus of politics and economics. Controlling the flow of capital and wealth within a consortium of banks and/or as a consortium of whales certainly does give you the power to control an entire economy.

Market Failures. In fact, in this case we already see the whales manipulating the crypto market quite frequently with their sell- and buy-walls, bear raids, pump-and-dump attacks, and other illicit strategies executed to manipulate the market. . . . These are market failures, not the mechanics of a free market. Market failures are fundamentally spawned by dysfunctional monetary and regulatory policies. In crypto markets, these market failures are the direct result of extremely high concentrations of crypto wealth in the hands of a few, which is the direct result of poorly executed ICOs and monetary policies.

The people who have debated against my position on this issue have ignored many aspects of economics and politics in the real world, including the past decade of crypto politics in the real world. I've already written extensively in many threads about these dynamics so I'm not going to repeat them anymore.

========== END EXCERPT ==========

Money Supply

Every economy's money supply must start at some amount and increase to another amount over time to accommodate real economic growth, but how much money is the *correct amount*? The answer to that question depends on many factors and each economy has different socioeconomic characteristics throughout its lifespan that require different amounts of money at different points in time to function effectively. For Gini, the optimal starting money supply is 1 billion Gini units and the optimal maximum money supply is 10 billion Gini units. Gini's initial money supply and sustainable growth rate can be visualized in the following chart.

Gini Money Supply Growth

Money Supply	
G12,000,000,000	
G10,000,000,000	
G8,000,000,000	
G6,000,000,000	
G4,000,000,000	
G2,000,000,000	
G	

1	7	13	19	25	31	37	43	49	55	61	67	73	79	85	91	97	103	109	115
2018		2019		2020		2021		2022		2023		2024		2025		2026		2027	

Months & Years

Why 10 Billion Max? The maximum money supply of cryptocurrency projects today is all over the map—ranging from about 11,000, to Bitcoin's 21 million, up to 1 trillion units in some obscure cryptocurrency projects. Whenever we have asked project teams, "Why did you choose x units for your max money supply?" their answers typically amounted to, "It just seemed like a good number." Regardless of their intentions, their logic isn't rooted in a deep understanding of historical money supplies in the real world. In contrast, at Gini we examined the money supply issue from several different perspectives to develop a technically sound, logically consistent, and sustainable Gini monetary system based on the following principles.

10 Billion is the Magic Number. During our research and analysis, we made several interesting observations. In particular:

- **Approximate Total Ounces of Gold on Earth: 10 Billion.**[245] Pegging the Gini money supply to Earth's gold supply creates a finite and useful upper limit, which is consistent with thousands of years of sound, precious metals-based money supply management throughout human history.[246] This is also one of the reasons that Gini is working with several hundred mining companies throughout the developing world to help them securely and efficiently manage their precious metals supply chains on the Gini Platform.

- **Approximate Maximum Human Population on Earth: 10 Billion.**[247] Earth's finite natural resources create an upper limit on human population growth, which is one of the most significant factors that impacts money supply growth for every human population.

- **Approximate USD Money Supply in 1900: 10 Billion.** We

---

245 This includes total estimated mined and un-mined supply. The World Gold Council provides a slightly lower estimate, but there are numerous other estimates that are much higher. So, this is a reasonable average. The precise figure can vary depending on whether Troy ounces or Imperial ounces are used in the measurement. Troy ounces are more common for precious metals.
246 Money and Gold | World Gold Council. (n.d.). Gold.org/about-gold/history-of-gold/money-and-gold
247 Live Science. (2011). How Many People Can Earth Support? Livescience.com/16493-people-planet-earth-support.html

spent a significant amount of time analyzing how the U.S. money supply was impacted by various events between 1900 and 2018, including the 1913 creation of the Federal Reserve, WWI in 1914–1918, WWII in 1939–1945, advent of the Bretton Woods System in 1944, Marshall Plan in 1948, Korean War in 1950–1953, Vietnam War in 1955–1973, termination of the Gold Exchange Standard in 1971, Saudi/OPEC agreement to price all their oil in USD in 1974, Iraq/Afghanistan War in 2003–present, and quantitative easing (QE1/QE2/QE3) associated with the 2008 Financial Crisis. Based on our analysis, we observed that the money supply in the year 1900 was the most *pristine* state of the U.S. monetary system. Specifically, that was the last point in time before the USG began its endless inflationary wars and monetary policy manipulation. The following chart reveals the relationship between U.S. money supply growth and population growth.

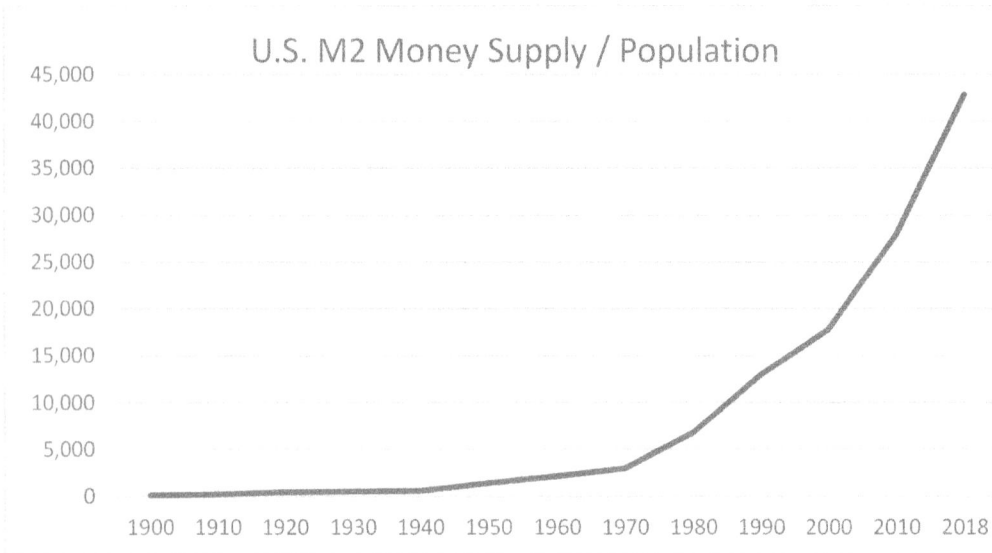

U.S. M2 Money Supply / Population

SOURCES: U.S. CENSUS, WORLD BANK, GINIFOUNDATION.ORG

The Gini Money Supply Is Designed to Preserve Gini Value.
One of the most important observations from our analysis is that a money supply must be designed to grow at a rate equal to or slower than

population growth to prevent currency debasement. For this reason, the Gini money supply is designed to grow slower than Gini stakeholder population growth. This ensures that the demand for Gini will be strong relative to the supply. Thus, the gradual increase in the Gini money supply to accommodate real-world commerce over the initial 10-year ecosystem-building period has a *dis-inflationary* effect. This prevents the value of Gini from being eroded by the natural initial supply growth required to create the Gini ecosystem.

Gini Currency Units Are Divisible to Accommodate All Future Needs. Just like 1 USD is divisible into 10 dimes or 20 nickels or 100 pennies without diminishing their total purchasing power, the Gini cryptocurrency is divisible into smaller units, too. As the Gini ecosystem expands, 10 billion base Gini units can be divided into 10^{28} Gini units without diminishing the purchasing power of the total Gini money supply. This will be increasingly necessary as the Internet of Things (IoT) and billions of autonomous distributed apps start producing trillions of micro-transactions every year.

Why Is Gini *Dis-Inflationary?* The population of stakeholders within the Gini ecosystem will grow over time. As the population grows, the Gini money supply automatically and naturally grows based on several mechanisms discussed in later sections. However, for now, it's important to understand that Gini's money supply growth is designed to be *dis-inflationary* (*not* deflationary) for three reasons:

(1) It ensures that the value of Gini currency is *never debased by monetary inflation.*

(2) It enables Gini to appreciate in value without discouraging real-world commerce.

(3) As every network of human users grows, there is a statistical reduction in network utilization per user *on average*. When that principle is applied to the Gini Network, it means there will be a statistical reduction in the total amount of Gini owned *on average* (per-capita) as the population grows. This is true for all networks as every network spawns *power users*, *newbies*, and every level of skill and participation in between, which is what creates the statistical dispersion of network/crypto utilization over time. The chart below illustrates this phenomenon.

Average Gini Supply per Stakeholder

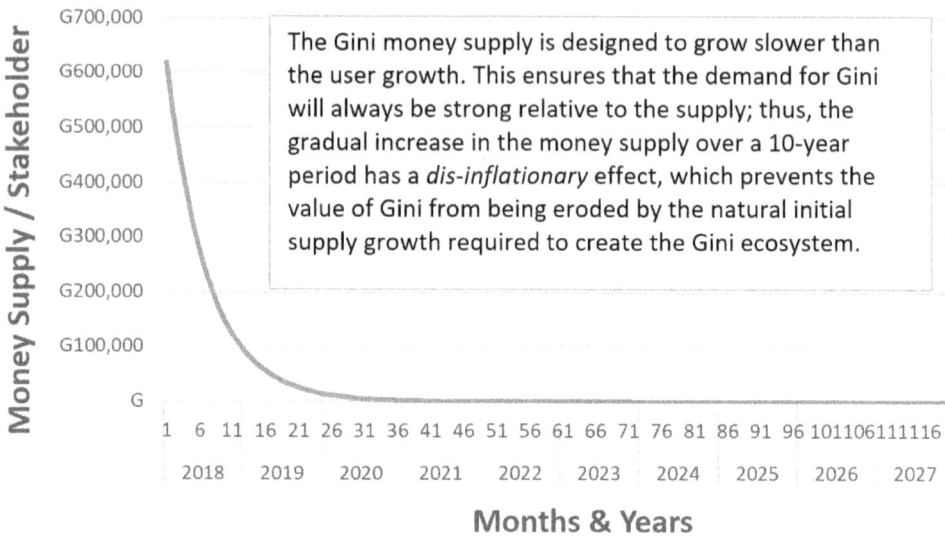

The Gini money supply is designed to grow slower than the user growth. This ensures that the demand for Gini will always be strong relative to the supply; thus, the gradual increase in the money supply over a 10-year period has a *dis-inflationary* effect, which prevents the value of Gini from being eroded by the natural initial supply growth required to create the Gini ecosystem.

Statistical Money Supply Dispersion *Is Not* the Same as Broad Wealth Distribution. Just because all networks tend to experience a reduction in per-capita resource utilization does not mean all cryptocurrency economies automatically produce a broad distribution of wealth. Many crypto project teams confuse statistical dispersion with actual wealth distribution. This is the same as confusing the concepts of per-capita wealth and median wealth. They are entirely different concepts. For example, if Bill Gates and two homeless people (each with only $1 in net worth) are in the same room, their *per-capita* (average) wealth would be about $31 billion in 2018! In contrast, their *median wealth* would be only $1.[248] This is another reason why crypto project teams that claim trickle-down economics will fix their wealth concentration problems over time are being disingenuous or naive.

---

248 See page 29 of the Global Governance Scorecard at Eanfar.org for a more detailed explanation of the impact of using per-capita vs. median wealth measurements.

The Law of Money-Value Creation

All Human Communities Must Inject Money Into Their Economies. In the fiat world, central banks create money out of thin air, then they inject that money into highly concentrated commercial banks, which then distribute that money to the rest of the population using the highly inflationary fractional reserve banking system. That's one way of injecting money into an economy, but it's not the most equitable, stable, or sustainable way to do it. When we have a liberated Gini ecosystem that doesn't depend on banks, we can bypass the banks and distribute the money directly to the citizens *as long as we respect the Law of Money-Value Creation.*

The Law of Money-Value Creation. In short, the "Law of Money-Value Creation" is a principle that I've developed over the years to help people understand the relationship between value creation and units of currency in a money supply. Formally, the law states: "Monetary inflation is the result of orphaned currency units in a money supply. To avoid inflation, each unit of currency must be linked to a unit of value in the real economy."[249] In other words, money cannot be sustainably created out of thin air like politicians and banks do today; rather, it must be created *and linked* with *actual value* that has been created in the *real human economy.* Specifically, money and value must be created and linked together (not necessarily at the same time); otherwise, orphaned currency units are created, which merely increases the money supply and creates monetary inflation without a corresponding increase in value within the real human economy. The image below illustrates this principle.

---

249 Do Air Drops Violate the Law of Money-Value Creation? No, see GiniFoundation.org/kb for a more detailed answer to this question.

EANFAR'S LAW OF MONEY-VALUE CREATION

Monetary inflation is the result of orphaned currency units in a money supply.
To avoid inflation, each unit of currency must be linked to a unit of value in the real economy.

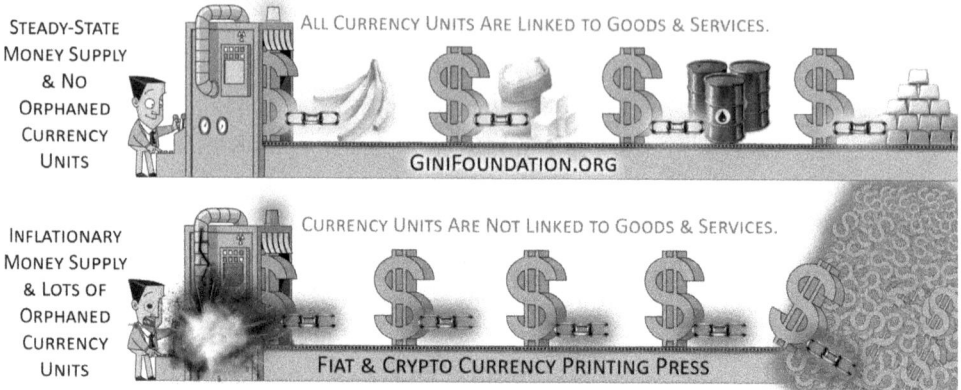

STEADY-STATE
MONEY SUPPLY
& NO
ORPHANED
CURRENCY
UNITS

ALL CURRENCY UNITS ARE LINKED TO GOODS & SERVICES.

GINIFOUNDATION.ORG

INFLATIONARY
MONEY SUPPLY
& LOTS OF
ORPHANED
CURRENCY
UNITS

CURRENCY UNITS ARE NOT LINKED TO GOODS & SERVICES.

FIAT & CRYPTO CURRENCY PRINTING PRESS

Money Supply Distribution

How Should Value Flow Throughout an Economy? This is an important question that every monetary policy should address. If a monetary policy is guided solely by the fantasy of the *pure free market*, value will inevitably concentrate in the hands of groups who can manipulate the laws and regulations that control the flow of value throughout an economy. Value can flow through many *value creation streams*, including: the production and sale of commercial products and services; the delivery of *nonprofit* products and services; informal volunteer work; social activities that bring communities together; stewardship activities that protect the community from harm; nonprofit scientific research and educational activities that enhance the performance and value of the entire ecosystem; among others.

A Narrow Perspective on Value Creation is Destroying Capitalism. Nearly all fiat and crypto monetary systems today fail to take into account the diversity of value creation streams that actually exists in the real world. Instead of intelligently designing a monetary system that takes into account a broad diversity of real-world value creation streams as *an intrinsic part* of the monetary system itself, nearly all fiat and crypto policymakers today put their faith in the fantasy of *pure free markets*; or, they use fiscal policy and taxpayer funds to arbitrarily fund projects linked to their biggest campaign donors.

In either case, their artificially narrow perception of the value creation and distribution process produces profoundly negative socioeconomic consequences, broken incentives, market failures, and numerous tragedy of the commons problems. These problems are destroying the broad wealth-generating capacity of capitalism, the ecological environment, and the social integrity of human societies in many countries today.

Free Market **Price Signals Ignore Many Costs.** When policymakers are in *free market mode*, they often assume that market price signals alone are sufficient to define and transmit value throughout an economy. However, the deliberately broken capitalism that the largest corporations and their political patrons in government have given us today completely ignores many costs to society that are caused by their artificial and myopic definition of *price signals*. In particular, the costs of environmental pollution and degradation, natural resource depletion, destruction of human capital (due to land squatters and other economic inefficiencies), structural unemployment caused by broken trade and labor policies, human deaths and disease caused by unnecessary poverty, malnutrition and pollution. . . . These are all real costs for which taxpayers pay, but the gigantic corporations that profit from producing these negative externalities don't pay any of these costs because they can dodge taxes and pass all costs on to consumers.

Banks Don't Suffer from the Inflation that They Create for Everybody Else. In addition to all the obvious broken incentives within the banking industry, the fiat *banking system* places central banks and highly concentrated commercial banks at the center of the money supply distribution process. This rewards bankers simply for being bankers by giving them access to newly created money before it *trickles down* to the rest of the population. This gives the banks and their shareholders more purchasing power because they can spend and make profits by lending out newly created money before its inflationary effects dilute the purchasing power of the money supply. Then when inflation strikes, they can simply increase their interest rates and fees to preserve their own wealth and purchasing power while the rest of humanity suffers.

The Fiat Banking System Is Not Aligned with Long-Term Economic Sustainability. Allowing banks to be at the center of money supply distribution gives their executives and shareholders enormous economic and political leverage over the entire economy because they can dictate how the wealth and capital flows throughout the entire ecosystem. Then they use their special access to newly created money to generate profits as quickly as possible, often from speculative land squatters and financial engineering scams. That means they have no economic incentive to fund long-term ecosystem-building activities that would strengthen the economy over the long-run.

Trickle-Down Crypto Monetary Policies Create the Same Problems. Merely dis-intermediating the banks is not enough. The money supply itself must be decentralized, too. This is why all major cryptocurrencies today are hopelessly broken. They create the illusion of decentralization by decentralizing their *physical* networks, but *their money supplies* are far more concentrated than any fiat economy today. Moreover, their crypto money supplies will always be far more concentrated than the money supplies in fiat economies because their project leaders don't have the economic incentive or ideological flexibility to implement meaningful ecosystem stability mechanisms. The primary power of an economy comes from the flow of currency, not its network topology. Whoever controls the currency controls the economy. Thus, all these super-concentrated cryptocurrencies today are nothing more than interesting science projects at best; at worst, they're new crypto tyrannies trying to replace the old fiat tyrannies.

Gini Value Streams

Gini's Value Stream System is the mechanism that prevents any stakeholder (or cartel of stakeholders) from completely dominating the Gini ecosystem. Fiat and crypto oligarchies exist throughout the world today because the fiat banking system and all other major cryptocurrencies are dominated and corrupted by a tiny number of *insiders* who use their financial wealth and political power to dominate those systems to serve their own interests. Their ability to dominate those systems is based on the structural design of those systems' monetary policies and their unsustainable and inequitable mechanisms of wealth creation and distribution.

Gini's Value Streams Are the Cure for Many Socioeconomic Problems. The systemic flaws in the fiat and other cryptocurrency systems today give insiders and super-wealthy entities within those ecosystems the power to block the general population from gaining any meaningful influence over their economic and governance decision-making processes. This has historically led to many toxic problems, including banking crises, devastating world wars, soul-crushing structural poverty, class warfare, social instability, corporatocracies, fascism, communism, violent revolutions, and many other socioeconomic problems. The Gini Value Stream System (GVSS) is the optimal solution for all these problems.

Broad Value Creation & Distribution is the Key to Equity, Stability & Sustainability. Like all fiat and crypto economies, the Gini ecosystem has incentive structures to encourage productive behavior and discourage destructive behavior. Additionally, like other cryptocurrencies, Gini stakeholders can be rewarded for mining and creating blockchain blocks. However, *unlike* other cryptocurrencies that only reward one category of stakeholders—miners or stakeholders who are already super-rich within those systems—the GVSS rewards multiple categories of stakeholders for contributing and adding value to the Gini ecosystem in several ways.

Synergistic Value Streams, Equity, Accountability & Sustainability. To more accurately reflect *and reward* the diversity of value creation streams in the real world, the GVSS ensures that value and purchasing power flow throughout the Gini ecosystem in an equitable and sustainable way. The GVSS is based on the synergistic principles of equity, accountability, and systemic sustainability. The pie chart illustrates how the Gini money supply naturally and

Gini Value Streams

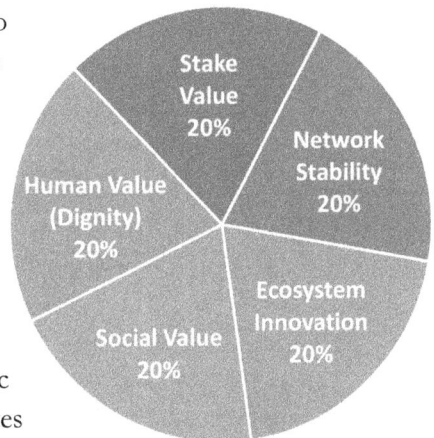

Stake Value 20%

Network Stability 20%

Human Value (Dignity) 20%

Ecosystem Innovation 20%

Social Value 20%

proportionally flows through each *value stream* within the Gini ecosystem based on five core areas of ecosystem value creation: stake value, network stability, ecosystem innovation, social value, and human value (dignity).

Gini Currency Exchange Payments vs. Commercial Payments vs. Treasury Payments. Like in any fiat economy, there are two ways to obtain Gini currency ("Gini"): purchase it or earn it. Purchasing and earning Gini can occur in the following three scenarios.

- **Currency Exchange Payments.** A Gini currency exchange payment can occur during an initial currency distribution event or a currency exchange transaction in Gini's decentralized market exchange. In these "currency exchange payment" cases, you pay in one currency and get Gini in exchange.
- **Commercial Payments.** Like any fiat economy, you can earn Gini by working directly for another stakeholder who pays you in Gini. You can also sell goods and services in exchange for Gini. These are all "commercial payments." These payments come from the existing circulating money supply.
- **Treasury Payments & the "Earnable New Money Supply."** *Unlike* fiat currencies and other cryptocurrencies, you can be paid in Gini *automatically* directly from the Gini Treasury for performing certain activities, which are classified into five "value streams" in the previous pie chart. These automated payments are "Treasury payments," which *create new money that is added to the circulating money supply.* (Remember, the circulating supply *never* exceeds 10 billion Gini.) The concept of "value streams" only applies to this newly created portion of the Gini money supply that is automatically paid directly from the Gini Treasury to stakeholders who have *earned* the payments in some *measurable way.* This is the "Earnable New Money Supply."

Earnable New Money Supply = Community Pool. From this point forward, whenever we talk about "monetary policy," "Treasury distributions," "blockchain block creation rewards," "Gini Community Rewards System" and any other type of automated payment directly from the Gini Treasury that is not a currency exchange or commercial payment as described in the first two bullet points above, we are talking about Gini Treasury payments from the Earnable New Money Supply, which are automatically governed by the Gini Monetary Policy. But because "Earnable New Money Supply" is tedious to remember, we call this the "Community Pool" to make it easier.

Injecting Money into the Gini Ecosystem via Value Streams. Now, we can turn our focus to the specific purpose of Gini's five core value streams. Based on 100% of the Community Pool at any point in time, each of the five core value streams represents 20% of the Community Pool. More specifically, each section represents 20% of the total Community Pool that can be distributed throughout the Gini ecosystem, in accordance with the Gini Monetary Policy.

The five core value streams represent a variety of activities, which enables Gini stakeholders to participate in the Gini ecosystem and earn Gini from the Community Pool based on the following criteria.

- **Stake Value Stream (20%).** This portion of the Community Pool flows to *all Gini stakeholders* on a *pro rata* basis, which is based on how much Gini they already have. It is intended to reward Gini stakeholders for holding their Gini, just like banks reward their depositors with interest income for holding their fiat currency. Giving stakeholders an economic incentive to hold onto their currency helps to increase the stability of the ecosystem and encourages savings and self-sufficiency in their daily lives. Just like a bank account, if they don't have any Gini in their account when this value stream is randomly triggered, then they don't receive any new Gini until they add more Gini to their account.[250]

- **Network Stability Value Stream (20%).** This portion of the Community Pool flows to stakeholders whose *nodes* are automatically and randomly selected to become Dynamic Guardian Nodes, which are similar to Bitcoin miners, but they

---

[250] This value stream is randomly triggered to prevent stakeholders from trying to predict the timing of the distributions, which would enable them to game the system.

are selected randomly and they don't require supercomputers to produce Gini blockchain blocks. Like miners on the Bitcoin Network, Dynamic Guardians are essential for maximizing the stability, speed, scalability, and sustainability of the Gini Network. This value stream is rewarded instantly as each new block is created and it's intended to provide an economic incentive for all stakeholders to keep their computers running to support the Gini Network. Any node that has the basic widely available technical specs will be automatically and randomly selected. (Dynamic Guardians are covered in more detail in the next chapter.)

- **Ecosystem Innovation Value Stream (20%).** This portion of the Community Pool flows to stakeholders based on several community-defined metrics that measure the value and impact that their technological innovations contribute to the Gini ecosystem. This includes any software tools, Web-based and distributed apps, core Gini Protocol improvements, hardware devices, and all other unique technology contributions that are made to the Gini ecosystem. Rewards from this value stream are automatically and democratically distributed by majority vote from the Gini Community.

- **Social Value Stream (20%).** This portion of the Community Pool flows to stakeholders based on how many people they bring into the Gini ecosystem. This is essentially an affiliate program, which rewards Gini stakeholders whenever somebody joins the Gini Network using their unique affiliate code. Every community needs sincere and enthusiastic ambassadors, including the Gini Community. Some stakeholders will work hard in this value stream like they would in any other job; so, the reward deposits for this value stream are distributed automatically on a predictable monthly schedule to reward them for their hard work.

- **Human Dignity Value Stream (20%).** This portion of the Community Pool flows randomly to all *human Gini stakeholders* on a *per-human* basis, which is based on recognizing each stakeholder's essential value and dignity as a human being. This may seem strange now, but in the future, many types of A.I. will also be stakeholders on the network; thus, we are creating an ecosystem that places a higher priority and value on human welfare over the welfare of A.I. stakeholders. For Gini accounts to be eligible for this value stream, a stakeholder's account must be verified as *human*. This is accomplished with well-established zero-knowledge proofs, which enable the Gini Network protocol to automatically confirm that an anonymous stakeholder's account is controlled by *a unique human* without leaking any of the stakeholder's private information to the Gini Foundation or anybody else. The entire process is completely secure and anonymous.[251] Important: *This is an opt-in feature; stakeholders do not need to verify their humanity to participate in the Gini ecosystem*, but Gini accounts that are not verified as human are not eligible for this Human Dignity Value Stream.

What Exactly Does "Flow to Stakeholders" Mean? In return for participating in one or more of the value streams defined above, Gini stakeholders earn a certain amount of Gini based on each stakeholder's current eligibility status for each value stream. The table below summarizes everything covered in this section so far and illustrates the mix of individual and ecosystem-wide incentives that are combined to achieve an optimal distribution of value, wealth, and power throughout the Gini ecosystem.

Value Contributed	Share of Pool	Reward Type	Eligibility Based On . . .	Timing of Reward Deposit
Stake Value	20%	Individual	Stakeholder's current account balance.	Random
Network Stability	20%	Individual	Random Dynamic Guardian selection.	On each new block creation.
Ecosystem Innovation	20%	Individual	Winning a community tech reward vote.	On winning community vote.
Social Value	20%	Individual	Being an effective Gini Ambassador.	Monthly
Human Value (Dignity)	20%	Ecosystem	Having a beating human heart.	Random
Total	100%			

Community-Driven Value Injection Process. The approach

---

251 For more about zero-knowledge proofs, see: Wikipedia.org/w/index.php?title=Zero-knowledge_proof

described above enables the entire Gini Community to *automatically* work together as a highly efficient team to inject value, Gini currency, and purchasing power throughout the entire ecosystem gradually and sustainably over time without violating the Law of Money-Value Creation. We call Gini's unique combination of technical systems, processes, and features a "Community-Driven Value Injection Process" (CVIP) because it creates the optimal balance between macroeconomic monetary system stability and microeconomic incentive structures to create the most equitable, stable, and sustainable economic ecosystem. In other words, the CVIP creates the optimal conditions for broad value creation and distribution throughout the Gini ecosystem.

Payments from the Community Pool Must Follow a Clear Monetary Policy. We have taken the time to clearly specify how Gini value streams and the corresponding Community Pool work because these value streams are rewarded *automatically and paid directly* by the Gini Treasury without any Gini Foundation intervention. This means the rules must be transparent and subjected to community review at all times in the spirit of true democracy. The rules that govern these automatic Treasury payments are based on the Gini Monetary Policy, which is the collection of principles and mechanisms that we have been describing throughout this Crypto Monetary Policy chapter.[252]

Stakeholder Distributions

The nearby pie chart illustrates a common *initial* money supply distribution used by many other cryptocurrencies today. However, Gini's supply distribution is very different, as you will see after the chart.

Existing Cryptocurrencies Do Not Produce Equitable Wealth Distributions. In the pie chart above, notice the "Remaining for Mining" piece of the pie. In most cryptocurrencies, most of that portion actually goes to a tiny number

Common Crypto Distribut

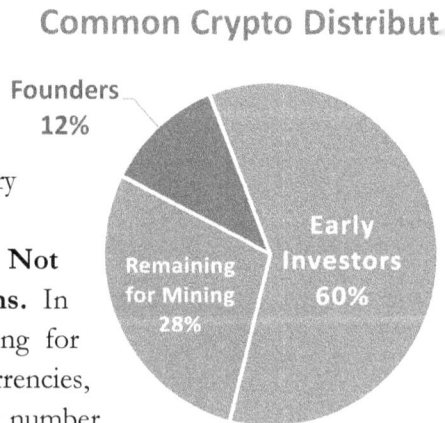

Founders 12%

Early Investors 60%

Remaining for Mining 28%

---

252 Gini's monetary policy principles can be viewed here: GiniFoundation.org/monetary-policy

of founders, early investors, and wealthy technology experts over time because their systems are designed to distribute the block creation rewards based on the distribution of wealth (stake) or hashing power (mining power). For Bitcoin and most other cryptocurrencies today, a tiny number of entities control over 90% of their mining power; and over 70% of their money supply is owned by a tiny cartel of founders, early investors, and miners. That means the *actual* amount of crypto wealth available to the general public over time will be significantly less than 25%. And without any ecosystem stability mechanisms, their distributions will inevitably become *more* concentrated over time. This makes them far worse than the economic tyranny in any fiat economy today.

Gini's Money Supply Produces an Equitable Wealth Distribution. In contrast to other cryptocurrencies, the Gini pie chart below illustrates how a far greater portion of the *initial* Gini money supply is allocated to the general public. Additionally, Gini's Ecosystem Stability Mechanism (ESM) *guarantees* that the long-run money supply available to the general public *and used to benefit* the general public *increases* to greater than 90% over time. (The actual amount will likely be closer to 95%.) This is true for three reasons:

(1) Gini's Community Rewards System and corresponding Value Streams automatically flow *equitably and sustainably* throughout the Gini ecosystem.

(2) The portion of the money supply owned and controlled by Gini's early founders/investors, ecosystem partners, and non-founding team shrinks as a percentage of the total money supply as the supply *naturally and sustainably* grows to the 10 billion maximum over a 10-year period.

Initial Gini Distribution

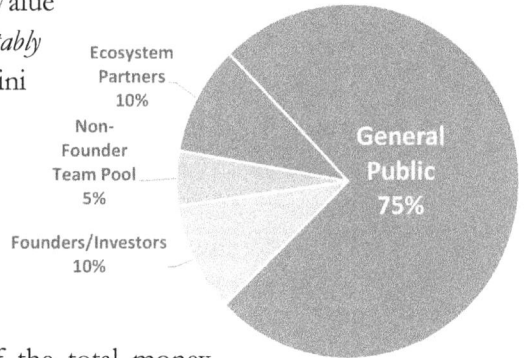

- Ecosystem Partners 10%
- Non-Founder Team Pool 5%
- Founders/Investors 10%
- General Public 75%

(3) A substantial amount of the Community Pool funds is distributed to worthwhile projects that benefit the entire Gini

ecosystem, not only a tiny number of miners, investors, and founders.

Gini Stakeholder Distributions. The nearby chart illustrates the specific distribution of the Gini money supply. Each piece of the Gini distribution pie is explained as follows:

- **General Public (75%).** This piece of the pie represents the portion of the Gini money supply that is initially allocated to the general public. "General public" means all stakeholders that are not founders/investors, Gini Foundation employees, or formal ecosystem partners. Gini's ecosystem stability mechanisms are specifically designed to ensure that the portion of the money supply allocated to the general public increases to *at least* 90% over time.

- **Ecosystem Partners (10%).** This piece of the pie represents thousands of third-party technical developers, nonprofit organizations (universities, colleges, research centers, think-tanks, etc.), and complementary organizations that contribute to the Gini ecosystem in meaningful ways. These Gini funds are specifically allocated to important technology, public relations, and legal defense projects that benefit the entire Gini ecosystem by building valuable services, tools, features, and legal defense strategies to improve the Gini ecosystem and protect the Gini Foundation from malicious attacks by banks and governments.[253] Collectively, Gini's Ecosystem Partners are crucial to the long-term success of Gini, which is why a meaningful portion of the initial money supply is allocated to this piece of the pie.

- **Non-Founder Team Pool (5%).** This piece of the pie represents the portion of the initial money supply allocated to the non-founder employees, independent contractors of the Gini Foundation, and general operations. Like any other nonprofit organization, Gini must be able to attract talented

---

253 See the "Blockchain Patent War" article to learn why "legal defense strategies" is important: GiniFoundation.org/kb/blockchain-patent-war-coming

employees and contractors; so, this portion of the initial money supply is dedicated to compensating all the employees and contractors who are necessary to manage Gini's infrastructure and operations.

- **Founders/Investors (10%).** This piece of the pie is allocated to Gini's founding team and early investors to compensate them for the tremendous financial and personal risks they are taking to build the Gini ecosystem. Given the status quo-resistant nature of the Gini Foundation, we are risking our careers, our existing companies, and even our personal safety. Thus, we hope readers appreciate the sacrifices we are making and agree that we deserve some compensation for all these risks. Also keep in mind that 10% is a much smaller portion of the pie than other crypto projects allocate to their founders/investors; so, we believe this allocation achieves a healthy balance and aligns the interests of the Gini founders/investors with the best interests of the entire Gini ecosystem.

Gini's Ecosystem Stability Mechanism. Recall that many other crypto project teams confuse per-capita wealth with median wealth, which means their claims that wealth and power concentration in their networks will decrease over time are not credible. Additionally, the monetary policies, systems, and blockchain protocol design of other cryptocurrencies guarantees that the wealth and power in their ecosystems will become *more* concentrated over time. In contrast, Gini's Community-Driven Value Injection Process (CVIP), Community Rewards System, and Value Streams guarantee *at the blockchain protocol level* that the concentration of wealth and power throughout the Gini ecosystem will decrease systematically and predictably. Collectively, we refer to all of these features as Gini's Ecosystem Stability Mechanism (ESM).

To visualize how the ESM causes the Gini money supply and wealth concentration to change over time, the following chart illustrates the estimated Gini money supply distribution in the year 2030.

A Guaranteed Equitable Wealth Distribution. The two Gini money supply scenarios illustrated in this chapter are guaranteed based on Gini's automated Ecosystem Stability Mechanism (ESM); so, there can be no favoritism or backdoors for a tiny number of whales or special interest groups to concentrate their wealth and power and take control of the Gini ecosystem. In contrast to the 80%-plus Gini Indexes of all major crypto-currencies today, the ESM guarantees that the Gini ecosystem will never have a Gini Index that exceeds 25% after the systematic CVIP process is complete. This is the optimal distribution of wealth throughout a population based on the highest quality of governance and quality of life outcomes achieved by the most politically stable and egalitarian countries on Earth today.[254]

Gini Distribution in 2030

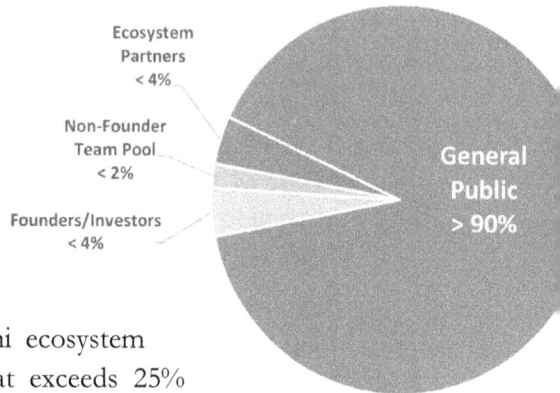

Ecosystem Partners < 4%

Non-Founder Team Pool < 2%

Founders/Investors < 4%

General Public > 90%

The Gini Difference. No other cryptocurrency has been created with as much careful attention to detail regarding the distribution of wealth and power within their ecosystems. Additionally, all the ecosystem benefits discussed throughout this chapter are achieved while simultaneously preserving all the benefits of free-market capitalism because nothing in the logic of the ESM prevents any stakeholder from engaging in any volume of free-market commercial exchange. In fact, anybody can earn a Gini fortune if they achieve the pinnacle of success by creating great products/services and selling them to millions of other Gini stakeholders within the Gini ecosystem. At the same time, the ESM still ensures that their wealth and power are *automatically and proactively* balanced within the ecosystem. This is one of the most significant benefits that differentiates Gini from all other fiat and crypto currencies today.

---

254 Eanfar, F. (2018). Global Governance Scorecard.

Key Points

- **Monetary Policy Has the Greatest Impact on Citizens and Crypto Stakeholders.** Broken fiat and crypto monetary systems destroy time=money=life and they produce unsustainable ecosystems. To prevent the Gini ecosystem from being destroyed by the same problems that plague other fiat and crypto currencies today, we have invested an extraordinary amount of time, energy, and resources into developing the Gini monetary system. This ensures that the Gini ecosystem is built on the strongest and most sustainable socioeconomic and technical foundation possible.

- **Clarity of Purpose.** The Gini monetary system is based on Gini's unique, sustainable and equitable economic philosophy, which guides all our technical development decisions. This gives all Gini stakeholders exceptional clarity of purpose and organizational integrity, which we have not seen in any other cryptocurrency project.

- **Gini Value Streams, CVIP, & ESM Enforce Long-Run Sustainability.** Gini's equitable and sustainable currency distribution mechanisms are based on Gini's innovative Value Streams System, Community-Driven Value Injection Process (CVIP), and Ecosystem Stability Mechanism (ESM). We don't expect stakeholders to remember all these components of the Gini architecture, but developing a familiarity with how the Gini systems work in general will give stakeholders a deeper appreciation for how and why the Gini ecosystem is different from all other fiat and crypto currencies today.

- Chapter 8 -
The Gini Technology

"It has become appallingly obvious that our technology
has exceeded our humanity." – Albert Einstein

During the late 1930s and early 1940s, a sprawling community of over 130,000 scientists, engineers, and logistical and administrative personnel were busy making history. They knew they were working on something extraordinary. They believed their work would change the world. Most of them believed they were liberating humanity from the tyranny of . . . fossil fuels. They believed they were building a new source of energy because that's what senior political and corporate officials in charge of the top-secret Manhattan Project told them. It was not until early August 1945 when they saw gruesome pictures of the charred bones, bubbling skin and oozing blood of hundreds of thousands of humans in two incinerated cities—Hiroshima and Nagasaki—that they all finally realized the truth: They built the deadliest weapon in human history.[255]

Technologies Are Often Hijacked. From energy technologies and the Internet; to medical and pharmaceutical technologies; robotics, A.I. and automobiles; among others, many important technologies have been hijacked by politicians and corporations to serve their narrow interests at the expense of the public interest.[256] They divert technologies from their optimal use in society toward deadly, socially destabilizing, and/or economically wasteful uses designed primarily to enrich a tiny number of gigantic corporations and their largest shareholders. Then, they use their

---

255 Dropping one atomic bomb on Japan might have been necessary, but dropping two certainly was not. If you're not sure about this, please read the article, "The Atomic Bomb & the Obliteration of Moral Authority": Eanfar.org/atomic-bomb-obliteration-moral-authority
256 For a detailed history of this problem, read: Internal Combustion: How Corporations and Governments Addicted the World to Oil and Derailed the Alternatives by Edwin Black.

wealth to bend political systems toward their own interests. *This is not capitalism* or the *free market* at work; it's corporate socialism, corporatocracy, and systemic corruption that enables the convergence of political and corporate interests to produce these suboptimal economic, social *and* technological outcomes.

The Biological Basis of Perceptual Imbalance. When humans are born, we have data-empty brains with only a few genetically inherited limbic system impulses. So, we have to fill our brains up with knowledge to be useful in human society. Scientists and technologists spend most of their lives studying and learning how to harness the power of Nature to build amazing technologies. R&D fills our brains with knowledge about narrowly defined technical domains. But *in general*, when we fill our brains with lots of technical knowledge and relatively little historical socioeconomic and geopolitical knowledge, it creates an imbalance of awareness. This imbalance can have many significant effects on our perception of the real world and how we perceive and respond to socioeconomic and geopolitical events.

Scientists & Engineers Hate Chaos. I have personally experienced this phenomenon because I periodically transition between highly technical activities (e.g., writing software code and managing technical projects), which are relatively structured and predictable, to analyzing and writing about socioeconomic and geopolitical events, which are often chaotic and much less structured and predictable. The instinct of scientists and engineers is to gravitate to things that are structured and predictable and to substantially ignore or resist things that are messy and unpredictable. The human brain subconsciously compels humans to avoid anything that makes us uncomfortable, but chaos and unpredictability are especially uncomfortable for engineers and scientists.

Math = Peace & Predictability. Every time we write an elegant algorithm or solve a hard technical problem, our brains release dopamine, which creates a feeling of pleasure. This reinforces our conscious and subconscious desire to focus on things that are relatively structured and predictable and to substantially ignore or resist everything else. When I'm alone writing code or studying how to implement some new technology, it's like a vacation from the messy world of socioeconomics and geopolitics. Science, engineering and any domain that is substantially based on mathematical principles is literally an entirely different world in which every input almost always has a predictable output, which creates predictable systems and outcomes. This structure and predictability create a perception of control, stability, peace, and tranquility in our lives; and it's wonderful.

Oblivion is the Price of Peace & Predictability. Now, we can understand the biological and psychological basis of several problems that afflict the technological world of blockchains and cryptocurrencies today. The structure and predictability in our technical work often creates an illusion—a false perception that we can build technical systems without taking into account the messy real world of socioeconomics and geopolitics. Additionally, the noise and propaganda about economics and global trade propagated by the mainstream media and gigantic corporations makes simplistic economic philosophies like Neoliberalism and Libertarianism attractive to humans who prefer simplicity over reality. Unfortunately, as explored more deeply in my previous nonpartisan book, *Broken Capitalism: This is How We Fix It*, those simplistic economic philosophies are grossly inadequate in a world where A.I. and the interests of gigantic corporations and politicians are converging to rapidly change virtually every aspect of human civilization.[257]

Why Are We Building the Gini Cryptocurrency?

Building a new cryptocurrency was not my first choice. In fact, I strongly urged several other major cryptocurrency project teams to take into account the problems described throughout this book and on our Gini website. In literally every case, they wrapped themselves in ideological dogma about capitalism and *free markets* and they conveniently ignored many real-world realities. So, after seeing these self-serving and ideologically-driven responses many times, I realized none of the other tech teams were willing or able to confront the obvious reality: Pure "free markets" are a fantasy and capitalism is

---

257 This is why we are building the Gini School of Economics at GiniFoundation.org.

broken; and one of the most significant reasons it is broken is because the monetary systems in virtually all fiat *and* crypto economies today are broken.

Innovation Without Real-World Impact is Meaningless. The Gini team and I have studied all the major cryptocurrencies and dozens of more obscure blockchain-based projects to digest the most innovative and important concepts in the cryptocurrency world today. As a result, we have come to an important conclusion: Many of them are interesting and some of them have made important contributions to the evolution of blockchain technology, but none of them provide the essential combination of features that humanity needs: A cryptocurrency that protects human rights, provides a sustainable monetary system, and maximizes the broad, wealth-generating potential of real-world commerce.

Focused on Practical, Real-World Goals. The Gini Platform is growing rapidly and some Gini technologies could evolve in meaningful ways by the time you read this book as we continuously improve Gini's systems and features. Additionally, the target audience for this book is the general public. As a result, we don't dwell on deep technical details here. For readers who want more technical detail, visit the "Technical" category of the Gini Knowledge Base at GiniFoundation.org. Some readers may also appreciate a more general introduction to blockchain technologies, which we provide at GiniFoundation.org/why-gini-cryptocurrency.

In the meantime, the rest of this chapter focuses on relatively high-level concepts to clearly explain why the Gini technical architecture is unique and essential to achieving our practical, real-world humanitarian goals. Since everything is interconnected, this chapter will also explore some of the social and economic implications of the technical architectural decisions we have made in contrast to the status quo in the crypto world today.

The Crypto Status Quo

Problems in the Cryptocurrency World Today. Several problems have emerged in recent years, which have generally caused the cryptocurrency industry to lose sight of some important principles. These principles have been lost for the following primary reasons:

- Feuds between competing cryptocurrency teams who want to dominate the cryptocurrency universe to feed their egos.
- FUD (fear, uncertainty, and doubt) propagated by investors who have a vested interest in their favorite crypto projects.
- Whales who use their financial power to manipulate crypto markets and the governance processes of crypto projects to maximize the return on their investments.
- Partisan politicians who spread FUD because they are afraid of losing control of their citizens and their tax-generating capacity.
- Gigantic banks and profit-seeking corporations that spread propaganda and FUD to perpetuate their existing dominance of commercial markets and industries.
- Ideological nonsense and noise about capitalism vs. collectivism, which causes many puritanical libertarians to ignore the realities of how socioeconomics and geopolitics influence the development of technologies that support economic and political systems in the real world.

Truly Sustainable Ecosystems Are Impossible with the Status Quo. The problems above exist for sociological and philosophical reasons, but they have distorted the technological development of most cryptocurrencies today. These problems result in flawed and unsustainable systems, which then become extremely complex because the crypto teams are forced to continuously cobble together patches and extensions to fix the fundamental problems in their architectures.

Unfortunately, these alleged *solutions* will never enable them to achieve all the most important goals of a cryptocurrency: sustainable monetary system, *meaningful* decentralization, *real* transaction privacy, immutable transaction history, *real* scalability, and *meaningful* viability in real-world commerce. To understand why this is true, let's review the fundamental purpose of a decentralized, distributed, public ledger.

The Purpose of a Decentralized Ledger. The primary purpose of a *decentralized* ledger is to prevent any single party or cartel from manipulating the ledger. Decentralization is essential to guarantee the privacy, transaction integrity, and economic freedom of the participants within an economic ecosystem. Yet, every major cryptocurrency project today has implemented proof-of-work and/or proof-of-stake systems that inevitably result in high concentrations of network hashing and/or staking power, which results in highly centralized networks. That makes them vulnerable to all of the following risks:

- Creating new money out of thin air, i.e., secretly expanding the crypto money supply.
- Editing the transaction history to reduce or increase the amount of money paid/received in a given transaction.
- Deleting illicit transactions to conceal a party's participation in a transaction.
- Creating fake illicit transactions under another party's name to frame or blackmail them, which could cause the innocent party to be blamed for crimes they didn't commit.
- Blocking or censoring citizens from engaging in *unacceptable transactions* for ideological, political or monopolistic reasons, which self-serving politicians and corporations often do.
- Many types of fraudulent transactions, which are intended to distort the truth about the provenance or purpose of transactions.
- And numerous other forms of ledger manipulation that are possible when the integrity of the ledger is vulnerable to high concentrations of coercive power on a blockchain network.

The Purpose of a Publicly Visible Ledger. The primary purpose of a *publicly visible* ledger is to enable objective third-parties to verify that certain transactions have occurred and to validate that those transactions match the transaction description claimed by the transacting parties. In other words, public visibility makes auditing the ledger possible, but the way virtually all public cryptocurrency ledgers are designed today is a disaster for human rights because they substantially destroy transaction privacy.

The Purpose of a Distributed Ledger. The primary purpose of a *distributed* ledger is to *distribute* the workload associated with maintaining a cryptocurrency system so that no single entity or cartel is responsible for all the costs associated with assuring the security, stability, and integrity of the system. A truly distributed workload also ensures that no single entity or cartel is able to gobble up all the transaction fees and rewards associated with supporting the system. Unfortunately, cartels and whales completely dominate all major cryptocurrencies today because their architectures and protocols do not sustainably distribute their workloads *and* rewards throughout their ecosystems.

What's Wrong with Existing Consensus Protocols?

The Proof-of-Work Consensus Protocol. Proof-of-Work (PoW) algorithms like the one used in Bitcoin's *protocol* (i.e., decentralized software) require each node on the network to solve a computationally-intensive math puzzle before it can add a new block to the blockchain. A node *proves* that it has earned the legitimate *right* to create the next block by presenting a solution to the math puzzle, which the Bitcoin protocol can easily verify mathematically. Upon verification, the block is accepted into the blockchain and the entity is rewarded with a *mining fee* in the form of a certain amount of bitcoins. Bitcoin was an important proof-of-concept that launched the cryptocurrency industry, but PoW has several significant drawbacks.

Pure Proof-of-Work Protocols Are Unsustainable. *Pure* PoW is the most commonly used consensus protocol in cryptocurrencies today. However, PoW-based blockchains are unsustainable because PoW unnecessarily wastes electricity and incentivizes users to waste money on expensive computing hardware and services. If Bitcoin was a country, it would be in the top-50 highest energy-consuming countries on Earth (#43 in the chart below) . . . and *Bitcoinistan's* energy consumption continues to grow rapidly.

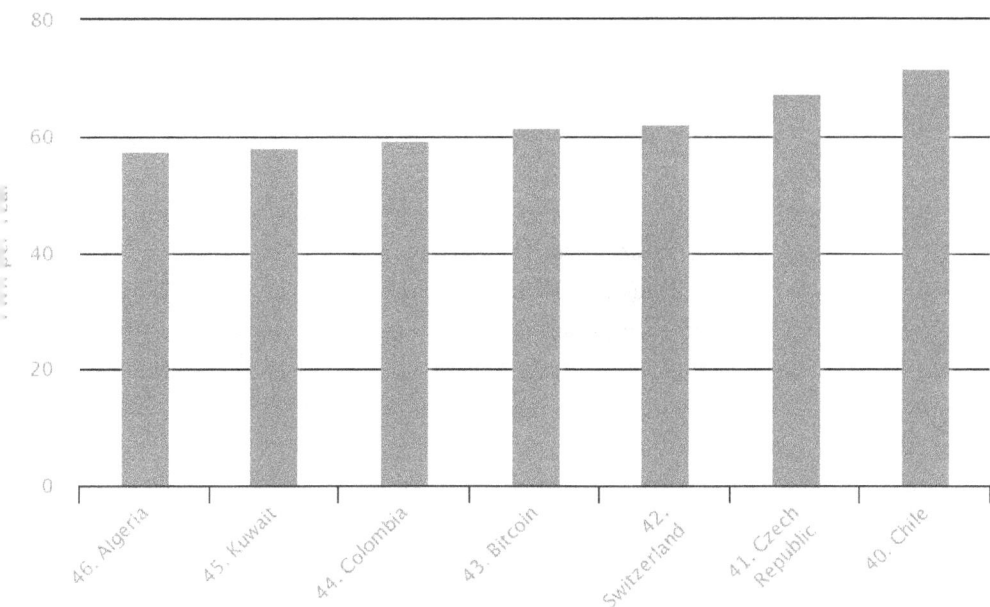

USD Billions of Wasted Resources. The chart above does not include the energy consumption associated with manufacturing USD billions worth of Bitcoin mining equipment or the energy consumption and equipment manufacturing associated with all other PoW-based blockchains. These expenses and byproducts waste finite natural resources and add no human value to the world beyond the profits they produce for a tiny number of companies; and those profits are primarily consumed by an even smaller number of the largest shareholders in those companies.

Proof-of-Work Cannot Scale. As the chart above illustrates, Bitcoin consumes more energy than most countries on Earth! In addition to the obvious environmental problems this creates, the cost of running the Bitcoin network will eventually exceed the value of the entire global economy. This results in transaction fees that already exceed the value of many products and services, which makes it impossible to use Bitcoin in real-world commerce.

Techno-Oligarchies. PoW-based blockchains typically result in a technological arms race, which concentrates wealth and economic power into the hands of a tiny number of mining groups that can afford to invest USD millions in expensive computer equipment and facilities. This inevitably creates a techno-oligarchy and a perpetual aristocracy of techno-elites who substantially dominate the entire ecosystem. This is clearly illustrated in the chart below.

In the chart below, only five entities control nearly 75% of the entire Bitcoin network; and only seven entities control 91% of the network. For anybody who truly cares about decentralization, this is intolerable. Given that trickle-down economics does not work, there's no rational reason to believe the oligarchy will ever voluntarily distribute their wealth and power throughout the Bitcoin ecosystem.

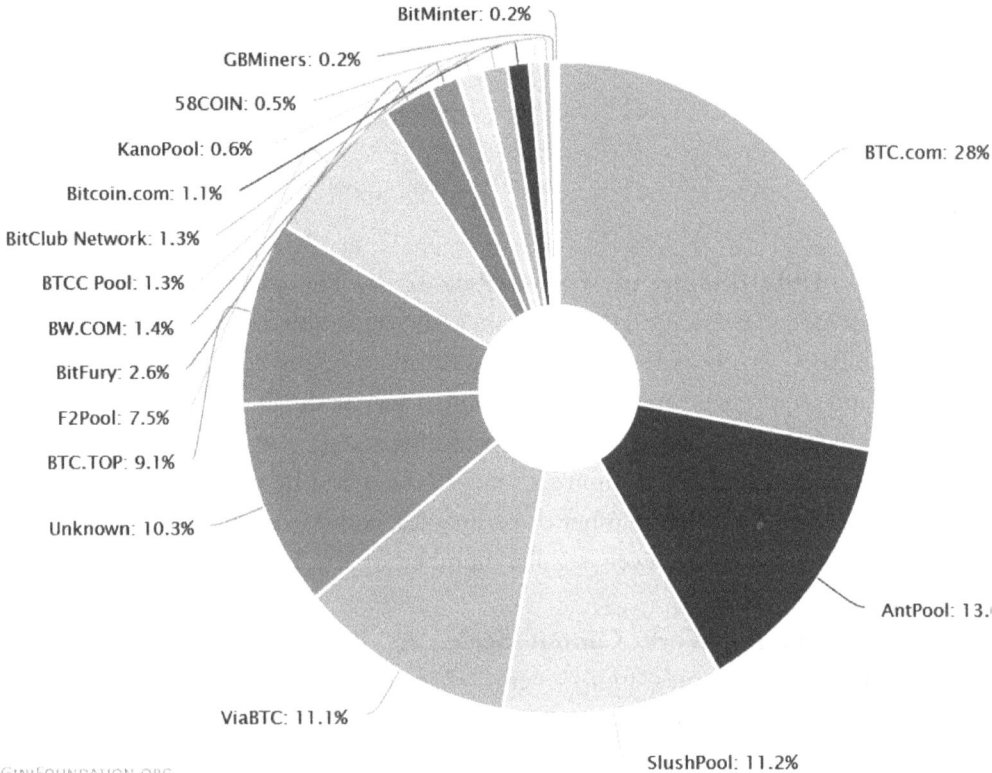

Proof-of-Stake Consensus Protocols. In contrast to PoW, Proof-of-*Stake* (PoS) consensus protocols grant power to stakeholders based on *stake*—essentially, the size of a stakeholder's crypto bank account. *Pure* PoS-based blockchains inevitably lead to high concentrations of wealth and power by rewarding whales simply for being whales. Advocates of pure PoS often embrace a puritanical form of libertarian ideology that can be boiled down to this: "Investors who risk their money should be given the most power. Whoever doesn't like it can sell their coins and leave. . . ." (That's a quote from somebody I debated in another cryptocurrency community.) If only human civilization, economics, and geopolitics were that simple.

What's the Point of Crypto Economies if They're More Concentrated than Fiat Economies? Both *pure* PoW and *pure* PoS inevitably lead to perpetual cryptocurrency oligarchies and *crypto-political dynasties* just like humanity suffers from in the fiat world. This creates a perpetual cat-and-mouse game with whales and crypto mining pools who inevitably use their power to prioritize their own interests above the interests of the broader ecosystem. This has profoundly negative consequences for any crypto or fiat ecosystem's long-term development and governance.

More Complexity Does Not Fix Anything. Many crypto teams think they can fix their problems by creating ever-more complex systems. For example, the Bitcoin community is trying to reduce their high transaction fees with new approaches like the Lightning Network. However, even if they could reduce the transaction fees, they will never be able to eliminate all the other problems because Bitcoin is already dominated by a tiny number of powerful special interest groups, just like the fiat-based economic and political systems are today. In other words, complexity doesn't change the fundamental reality: Any crypto or fiat ecosystem that is not . . .

(1) launched on a solid monetary system foundation of broadly distributed wealth and power; and

(2) built with *proactive and automated* ecosystem sustainability mechanisms that ensure the broad and sustainable long-term creation and distribution of wealth and power;

. . . will never escape the trap of concentrated wealth and power without a violent revolution.

Crypto-Oligarchies Lead to Revolutions, Too. Violent

revolutions have occurred frequently in human history in response to high concentrations of wealth and power in the fiat world. In cryptocurrency projects, "violent revolution" means the community and software code fork into two or more communities and code repositories, which often do not survive the divorce. Gini's technological architecture is the result of a philosophical revolution against the puritanical and unsustainable libertarian ideology that dominates virtually all cryptocurrency projects on Earth today.[258]

The Gini Architecture

In chapter six, we discussed the history of cryptocurrencies and blockchains in general. Now, let's cover the features of Gini's particular database structure, which we call a "BlockGrid."

What is the Gini BlockGrid? The Gini BlockGrid serves the same basic purpose as a blockchain, but it's much faster, more secure and more efficient. The BlockGrid is a hybrid database structure that possesses the most useful properties of a blockchain and some useful properties of a Directed Acyclic Graph (DAG). A DAG is a class of data storage structures that stores data within a parallel, node-based graph structure that is much more efficient than the typical singly-linked-list blockchain structures used by Bitcoin and most other cryptocurrencies.

However, unlike other DAG's, Gini's BlockGrid is decentralized and distributed as an interconnected *grid of private blockchains*, which dramatically increases the security, privacy, speed and throughput of the Gini cryptocurrency compared to previous generation cryptocurrencies. The Gini BlockGrid technically isolates each user account from all other user accounts, which protects the privacy of each account, but each stakeholder still has the ability to transact directly *and anonymously* with all other stakeholders with nearly instantaneous transaction confirmations.

A Provably Secure BlockGrid. The Gini BlockGrid is designed to be a *provably secure* system, which means it is possible to employ

---

258 Recall every member of the Gini team was previously a libertarian; so, we've seen the world from both perspectives. See the Gini School of Economics at GiniFoundation.org for context.

mathematical analysis of Gini systems to provide quantifiable assurances that the source code performs as expected within specific statistical thresholds of security and safety. This is the same principle used by aerospace companies to ensure that their airplanes, spacecraft, and other mission-critical systems perform as expected. This is important for a cryptocurrency because it dramatically reduces bugs and security weaknesses, which many other cryptocurrencies have suffered from in recent years.

Multi-Layer Architecture. Gini is designed with a multi-layer architecture that separates the Gini Settlement Layer (GSL) from the Gini Computation Layer (GCL). The settlement layer is where all the private currency and account balance data resides; whereas, the computation layer is where all the smart contract logic resides. This architecture increases the security and resilience of the Gini Network and enables organizations to optimize and configure Gini to suit their specific needs.

GiniScript Smart Contracts. The GCL includes a simple scripting language called GiniScript, which gives Gini stakeholders a user-friendly way to create automated sale and purchase contracts, loan contracts, insurance coverage contracts, escrow contracts, service contracts, and many others. These smart contracts enable relatively non-technical stakeholders to *automatically* hold their counterparties accountable in real-world commerce.

The Gini Account Center. All cryptocurrencies depend on an account management software program (aka *wallet*). The decentralized Gini Account Center enables Gini stakeholders to hold their cryptocurrencies safely and securely just like a bank account enables you to hold your fiat currency. But unlike a bank account, all Gini accounts are decentralized, *anonymous and private by default*.[259] Even the Gini Foundation cannot see any stakeholder's private Gini account nor their personal identity. Gini's stakeholder account software is designed to be simple to use for anybody who is familiar with the typical online banking

---

259 Gini's privacy features are continuously evolving to integrate industry best practices and to achieve the optimal balance between privacy and performance.

software used by many banks today.[260]

High-Reliability Gini Code. From the perspective of organizations that intend to use the Gini BlockGrid for their own mission-critical applications, one of the most attractive benefits of Gini is the way the software code is created. The core Gini components are produced with a *functional* programming language called Haskell, which has unique features that make it up to 80% more time- and cost-efficient than other *imperative* (aka, *procedural* or *object-oriented*) languages like C++, C#, Java, and most other languages. Haskell dramatically reduces source code complexity, which dramatically reduces the number of bugs and security holes throughout the software's lifecycle. This results in much more elegant and maintainable code, which results in substantially more reliable and cost-effective systems.[261]

Integration with Thousands of Merchants. Since we built Authorize.Net (the world's first and largest online payment gateway) and then AngelPay (the world's first and largest nonprofit payment processor), there are hundreds of thousands of merchants, collectively processing many millions of transactions every day, using the payment technologies that the Gini team has created. As we build and deploy the Gini technology, we are gradually integrating all our existing and future merchants into the Gini ecosystem over time.

The Gini Trust Protocol

We are all familiar with the concept of "trust" in human relationships, but trust also applies to institutions and technical systems. Trust is one of the most significant factors that determines whether humans will use a particular technology or not. In fact, when it comes to cryptocurrencies, the concept of trust is one of the most hotly debated topics today.

The Religion of "Trust-less" Blockchains. In the beginning of every crypto-fan's journey into the world of decentralized systems, we are introduced to the concept of "trust-less" systems. The concept of

---

260 Visit GiniFoundation.org to view the latest software development details.
261 Learn more about Haskell here: GiniFoundation.org/kb/haskell

building a system that doesn't depend *exclusively* on trust in any *particular* human or entity is crucial to building high-integrity systems and institutions. However, the puritanical ideology at the heart of the cryptocurrency world today has transformed this concept into a veritable religion. The mantra of this religion is "trust nothing." It's an inspiring mantra for those of us who are sick of seeing our economic and political systems destroyed by supposedly *trusted* corporations and politicians, but the people who embrace this mantra often ignore many technical realities associated with how scalable, decentralized systems work in the *real world.*

Decentralization in the Real World. Anybody can read a dictionary or textbook to memorize the academic definition of "decentralized," but in the real world, things are much more complicated. To understand why the puritanical "trust nothing" religion is problematic, we need to understand a few core principles about system design in the real world.

- The concepts of "trustworthiness" and "trusted" are not synonymous.
- Something that is trusted for a short period of time doesn't mean it must be trusted *forever.*
- A system without a way to allow stakeholders (humans or non-human nodes) to *earn* certain levels of trust will never be truly decentralized because it will never be truly scalable. Why? Because the system's administrators will never be able to effectively perform all the work required to maintain any large, decentralized system without being able to delegate certain activities to *trustworthy* ecosystem participants. Without *meaningful*, broad-based delegation to *dynamically trustworthy* ecosystem participants, the system will inevitably become highly concentrated like the crypto and fiat markets are today. Thus . . .
- *Trustworthy delegation* is the most important principle in building any scalable *and* sustainable decentralized system.

Effective Delegation is Based on a *Continuum* of Trust, Not Binary Decisions. Binary decision-making is intoxicating because it *seems* to reduce the complex universe of choices to simple all-or-nothing, black-or-white, on-or-off, right-or-wrong decisions. I know from first-hand experience that many software programmers, engineers, and mathematicians get frustrated when unpredictable human nature craps all over their rationally designed models and systems.

The Real World is Not Binary. The real world of sociology, economics, and geopolitics rarely conforms to binary logic. The real world is usually messy and requires much more nuance than any binary decision tree can handle. Thus, anything at the intersection of human nature, economics, and geopolitics (e.g., cryptocurrencies) can never be based on a "trust nothing" ideology. In the real world, every system operates based on *a continuum of trust*, not a "trust or don't trust" binary decision.

The Gini Trust Protocol. The concept of a "consensus protocol" implies *trust in the system that facilitates consensus*; and that trust must be based on *something*. This is another reason why the "trust nothing" religion is not realistic from the perspective of anybody who has real-world experience in business, geopolitics, or complex human-system interface design. In reality, a "consensus protocol" alone is not enough; blockchain teams need to adopt a *continuum of trust* if they want to build scalable, sustainable, decentralized systems. For this reason, and to reflect the reality of the real world, we call the Gini BlockGrid protocol the "Gini Trust Protocol," not merely a "consensus protocol."

Dynamic Proof-of-Commitment. The Gini Trust Protocol is substantially based on Gini's unique Dynamic Proof-of-Commitment (DPoC) algorithm, which automatically and dynamically selects and rewards high-quality Gini accounts/nodes and gives them *Dynamic Guardian* status on the network. Being a Dynamic Guardian is the highest and most rewarding privilege in the Gini ecosystem. Thus, there are strict requirements to be eligible for Guardian status. Guardian status must be earned and automatically and *continuously* verified based on several factors. This ensures the integrity of all Dynamic Guardians operating on the network, which ensures the integrity of the entire Gini ecosystem.[262]

---

262 More about DPoC & Dynamic Guardians here: GiniFoundation.org/kb/gini-trust-protocol

Gini Prioritizes Consistency Over Instant Availability. In Computer Science, there is a principle called the CAP Theorem, which states that during a network attack (e.g., a DDoS attack) it's technically impossible for a distributed system to simultaneously provide more than two out of three guarantees: consistency, availability, and partition tolerance. Because financial transactions often include many dependencies on other transactions, trying to process transactions in real-time that are being disrupted by a large-scale attack could result in a cascade of out-of-sync data, which can then be exploited in various ways by bad actors.

Thus, consistency and partition tolerance are Gini's highest priorities. That means, in the hypothetical case when the Gini Network is under attack, the Dynamic Guardian nodes stop validating transactions (i.e., they reduce service availability) and they alert all other nodes that no more transactions will be processed until the attack is over. Prioritizing consistency and partition tolerance over availability is the only way to guarantee that stakeholder funds are protected during a network attack.

In Summary: The Gini Trust Protocol, Dynamic Proof-of-Commitment, Dynamic Guardian nodes, and the Gini BlockGrid architecture are all unique and important innovations that enable us to achieve Gini's humanitarian goals securely and efficiently. These features also enable Gini to achieve strong public and private ledger integrity, which automatically results in strong macro money supply integrity. Collectively, we believe these features are compelling reasons to participate in the Gini ecosystem.

True Privacy

Gini's Commitment to Transaction Privacy. If you have watched the Gini documentary film on the Gini website and read the previous chapters of this book, it should be clear by now that transaction privacy *is a human right*. Without transaction privacy, the human mind feels trapped and ordinary events become uncomfortable, anxiety-inducing experiences that undermine the natural interaction with people and our environment. There is no reasonable justification to give up your liberty and privacy to trillions of tiny cryptocurrency spies, infiltrating your life 24/7/365.

Gini enables you to engage in legal commerce with whomever you wish without fear of others judging you or tracking your every move. There are several ways to achieve privacy on a cryptocurrency network. Some approaches are more computationally efficient and effective than others. Here we summarize Gini's technical approach and why it's objectively superior to the approaches used by other cryptocurrencies today.

How Does Gini Compare to Other Privacy Coins? Cryptocurrencies that emphasize privacy are often called "privacy coins." They are usually based on the CryptoNote protocol. They protect privacy with computationally complex mechanisms called ring-signatures to protect the sender's address, stealth addresses to conceal the receiver, and "ring confidential transactions" (RingCT), which obscures the transaction amount. Additionally, some privacy coins use a technology called zk-SNARKs (zero-knowledge proofs), which can provide more anonymity than the CryptoNote-based coins. In general, these cryptocurrencies usually do provide solid privacy protection, but they impose several trade-offs.[263]

- Their computational complexity makes them extremely slow—even slower than Bitcoin's seven transactions per second. This is primarily because they've bolted their privacy features on top of the Bitcoin code base, which was only designed for anonymous transactions, *not truly private transactions*. But anonymity is very different from true privacy; so, it's often relatively easy to link Bitcoin transactions together to reveal the identity of the transacting parties. This is a risk that applies to all cryptocurrencies that are substantially based on the Bitcoin source code and others that have adopted Bitcoin's non-private anonymity model.
- Their complex architectures are more vulnerable to bugs and hacks than Gini's architecture because Gini is built with the

---

263 The information in this section is accurate as of early 2018.

Haskell programming language, homomorphic encryption, and true transaction privacy at its core. This means Gini's architecture is simpler, which enables the Gini Trust Protocol to process transactions much faster with less vulnerability to bugs and hacks.

- Other cryptocurrency transactions are not private by default; users must turn on privacy for each transaction. In contrast, all Gini transactions are private by default.[264]

- Other cryptocurrencies based on ZK-Snarks require a trusted setup process, which makes it difficult for many people to believe that it has not been set up with backdoors or built-in privileges that could secretly benefit the founders of those other cryptocurrencies. For example, they could arbitrarily create new crypto-money for themselves and nobody would be able to detect it. In contrast, there's no secret setup process for Gini's privacy features. Gini is designed to be private by default so that Gini stakeholders can control the privacy of their transactions at all times.

- There is no way to audit the money supply of those cryptocurrencies. In contrast, Gini's money supply is audited in real-time and anybody can verify that no new Gini money has been created out of thin air at any time.

- There is no way for authorized parties to easily audit individual transactions with those other privacy coins. In contrast, Gini is designed with unique auditing features that enable stakeholders to allow third-parties *that they authorize* to audit their transactions if they desire. This is useful for accounting audits and organizations that have government-mandated financial reporting requirements.

- Some privacy coins charge extra fees for their private transactions. Gini does not.

- Some privacy coins are based on permanent "masternodes,"

---

[264] Depending on when you read this book, the privacy features might still be in development. In the beta version of Gini, transactions are not private by default because we want to demonstrate the other core Gini features first, which are easier for most people to see and appreciate.

which require over USD 500,000 (as of April 2018) to purchase. That means only a small number of wealthy crypto whales have meaningful opportunities to operate those masternodes. Of course, this is completely opposite to how Gini works, as our Gini Trust Protocol and Value Streams and monetary system pages at GiniFoundation.org illustrate.

- With certain other privacy coins, hundreds or thousands of masternodes *and their corresponding cryptocurrency supply* can be secretly owned or controlled by the same entity because there are no ecosystem sustainability mechanisms to prevent this. Again, this is completely opposite to Gini's architecture and philosophy.

- All the early adopters in those cryptocurrencies purchased their masternodes when it was cheap, giving them perpetual aristocratic control over their crypto ecosystems. Given the hyper-concentration in all other crypto markets today, it's reasonable to assume that their networks are substantially and secretly controlled by a small number of entrenched masternode operators with overwhelming economic and political power over their ecosystems. This outcome is no different than all the other highly concentrated cryptocurrencies today. In contrast, Gini's architecture and philosophy are fundamentally designed to prevent this, as illustrated throughout this book and on the Gini website.

Homomorphic Encryption. Homomorphic Encryption is the most significant innovation in the field of cryptography in many years. Homomorphically encrypted programs never need to decrypt their inputs; so, they can be safely executed by an untrusted party without ever revealing their inputs and internal state to that party. This has many applications and profound, positive implications for securely outsourcing private computations, cloud-based systems and distributed, decentralized networks.

Homomorphic encryption enables Gini Guardian nodes to perform computations on encrypted data without ever needing to decrypt the data. We use homomorphic encryption to dramatically reduce the algorithmic complexity of Gini's privacy features, which dramatically increases the Gini Network's speed and throughput compared to other cryptocurrencies. This is how we preserve the privacy of the data objects that are processed by Gini's Guardian nodes while ensuring that the performance of the Gini Trust Protocol is still very fast.[265]

To be clear, we have a lot of respect for those other cryptocurrency projects and any project that takes transaction privacy seriously. However, Gini is substantially different on many levels and we feel it's appropriate to clearly explain these differences so that Gini is never incorrectly characterized as "just another privacy coin." Gini is much more than "just another privacy coin."

Ecosystem Stability Mechanism

No cryptocurrency or fiat economy is sustainable unless specific mechanisms are engineered into their architectural DNA to create gradual resistance against economic cannibals and unsustainably high concentrations of wealth and power.[266] For this reason, one of the most important components of the Gini architecture is Gini's automated Ecosystem Stability Mechanism (ESM). We briefly described the ESM in the previous chapter. Here, we dive a bit deeper into what the ESM actually does to ensure that the Gini ecosystem is as equitable, stable, and sustainable as possible. The ESM performs the following critical functions:

1. **Value Streams.** The ESM's Value Streams System ensures that all ecosystem participants are rewarded for their valuable contributions to the Gini ecosystem. As discussed in the previous chapter, this specifically occurs via Gini's five core value streams: Stake Value, Network Stability, Innovation Value, Social Value, Human Dignity Value.

2. **Dynamic Transaction Fee System.** The ESM ensures that all

---

265 Learn more about Homomorphic Encryption here:
Wikipedia.org/wiki/Homomorphic_encryption
266 Learn more about economic cannibals here: Eanfar.org/transnational-economic-cannibals

transaction fees are automatically reduced as the Gini currency value appreciates against other major fiat currencies. This ensures that stakeholders are not penalized for using Gini in real-world commerce just because their Gini currency appreciates.[267]

3. **Treasury Funding.** The ESM provides the minimum level of Gini funds to the publicly auditable Gini Treasury. This ensures that the nonprofit Gini Foundation is always solvent and sufficiently staffed to manage the Gini Network and effectively build new features that are requested by the Gini Community. The Treasury is also the primary funding source for the Gini Community Pool, as described in the previous chapter.

4. **Gini Index Sentry.** The ESM's Gini Index Sentry monitors the Gini Index of the Gini ecosystem and automatically and gradually creates structural resistance against high concentrations of Gini wealth to prevent any whale or cartel of whales from dominating the Gini ecosystem. (More on this later.)

5. **Market Exchange Sentry.** This automated sentry monitors the Gini Decentralized Exchange, watches for malicious and/or abusive market behavior, and anonymously throttles the transaction speed of Gini accounts that engage in such behavior, which protects the market from harm.

Now, let's discuss specifically how and why the ESM performs the functions summarized above.

Value Streams. Like the Bitcoin Network and all other serious cryptocurrencies and banks in the fiat world, the Gini Network must charge some kind of transaction fee to prevent malicious attacks and to be economically sustainable. However, unlike other cryptocurrencies and the fiat banking system, instead of distributing the fees to only a tiny number of Bitcoin miners or the largest shareholders of gigantic banks

---

267 Transaction fees on the Gini Network are lower than other major cryptocurrencies. The fees exist primarily to prevent malicious attacks on the network and to ensure ecosystem sustainability. Learn more about Gini transaction fees here: GiniFoundation.org/kb/gini-transaction-fees

and corporations, Gini's ESM distributes the fees more equitably throughout the Gini ecosystem.

Given the indisputable fact that a sustainable community is required to create sustainable value in any ecosystem, the Value Streams System is the most equitable, sustainable, and rational way to manage a money supply. Thus, the ESM executes all the Value Streams functions, which automatically, equitably and sustainably distribute wealth and purchasing power throughout the ecosystem according to the Value Streams parameters and Gini's sustainable monetary system principles discussed in the previous chapter.

Gini Index Sentry. Recall the Gini Index is the gold standard in Economics, which is used by the IMF, World Bank, many NGOs, and virtually all governments worldwide to measure the concentration of wealth within each country's population. The Gini Index enables us to measure the concentration of wealth in the Gini ecosystem, too, but we do something more useful with it: Enforce money supply best practices by ensuring that the Gini ecosystem's Gini Index never exceeds 25% for too long.

To accomplish this, whenever the Gini Index Sentry detects that the Gini ecosystem has exceeded the 25% threshold, it automatically triggers a *very gradual* Ecosystem Stability Fee (ESF) on any Gini account that is in the top-1% of Gini currency holders. This ESF is a fraction of 1%; so, it's very small but still enough to gradually reduce the systemic risk over time. The ESF continues to be automatically and gradually collected from the top-1% of all Gini accounts until the Gini Index falls below 25% again.

Is the Ecosystem Stability Fee a Wealth Tax? No, it's an *ecosystem stability fee*, which is a form of community-funded protection against the kind of malicious and reckless whales who have destabilized and destroyed many fiat and crypto economies and nations throughout the history of capitalism.[268] If any Gini stakeholder accumulates such a large Gini fortune that they push the Gini Index above 25%, that means they have become a systemic risk to the entire Gini ecosystem. Thus, the top-

---

268 Visit the Gini Book List to read many scholarly, nonpartisan books that confirm this reality: GiniFoundation.org/kb/gini-book-list

1% of Gini stakeholders have a shared responsibility to contribute the very small ESF to protect the ecosystem from potential instability and harm until the Gini Index falls back below 25%.

The ESF Codifies, Systematizes, and Enforces Ecosystem Stability and Cooperation. The tiny ESF reduces the incentive for whales in the Gini Community to act selfishly. It also helps to reduce the risk of all other stakeholders suddenly losing their purchasing power due to a few disruptive whales thrashing around the Gini ecosystem.[269] These systemic features also serve as gentle reminders about Gini's purpose: Protecting human rights; sustainable and equitable monetary system; and maximizing the broad-based, wealth-generating potential of real-world commerce. The ESF is a rational, economically viable, and justifiable trade-off that helps to achieve those goals, which are essential to ensure the long-run viability of any socioeconomic ecosystem.

Will the ESF Scare Away Some Whales? Possibly, but the foundation of every stable economy is a strong middle class, not whales. The ESF is a tiny *and temporary* fee that does not significantly impact the fortune or quality of life of any whale, but the ESF is just enough to incentivize Gini whales to behave responsibly or find another crypto home. Anybody who is rich enough to be a whale can invest their wealth anywhere they want, but if they want to participate in the Gini ecosystem, they need to respect the Gini Community's unique commitment to creating the most stable, equitable, and sustainable ecosystem. The trade-off for these benefits is some restrictions on whales. We welcome whales and minnows alike, but we understand Gini may not appeal to all whales. That's OK.

Market Exchange Sentry. The Market Exchange Sentry monitors and prevents abuses in the Gini Decentralized Exchange. This topic is a bit more technical; so, if you want to learn more, please visit the Gini Knowledge Base at GiniFoundation.org and search for "Decentralized Exchange."

Where Do the ESF Funds Go? The ESF funds are automatically deposited into the publicly auditable Gini Treasury account on a daily

---

269 See vivid examples of this problem: GiniFoundation.org/gini-decentralized-cryptocurrency-exchange

basis until the Gini Index falls back below 25%. Like all Gini Treasury funds, the ESF funds are used to recycle wealth and purchasing power back into the ecosystem. Unlike the secret and opaque accounts of most public and private organizations on Earth today, all Gini Treasury fund balances will be published publicly at the GiniFoundation.org website after the Gini Treasury is fully activated.

Auditing Gini Transactions

Before we focus on the relatively technical details associated with auditing Gini transactions, let's review a few important principles to appreciate why Gini's BlockGrid and privacy features work the way they do.

All Gini Transactions Are Private by Default. Regarding the transactions of *all private stakeholders* throughout the Gini ecosystem, there is only one thing that the *general public* has a legitimate need to verify: No new money is created out of thin air from any transaction. *That's it.* All your other private data, including the identities of the transacting private parties, your private payment addresses, the amount of your private transactions, the purpose of your private transactions, your private IP addresses, and all other private transaction details are nobody's business. If you want to reveal your private information to a merchant or stakeholder, *that's your choice*, but your anonymity *and* privacy should be protected *by default* in all other cases.

Block Explorers Violate the Human Right to Privacy. At this time, Gini does not provide a "block explorer" or any other way to *randomly spy* on your transactions like the block explorer on the Bitcoin network and many other cryptocurrency networks. In fact, a block explorer is a naive and terrible violation of the human right to privacy. As CPUs, GPUs and A.I. become more powerful, it will become even easier than it is today for oppressive governments and corporations to map entire cryptocurrency networks and trace publicly visible crypto transactions directly to their owners.

Block Explorers Are a Dictator's Best Weapon Against Human Liberty. Without truly private crypto transactions, there is no way to purchase the goods and services that are necessary to launch any meaningful protest or resistance against an oppressive government. This has already happened in China, North Korea, several GCC nations, and it's guaranteed to happen in many more countries as global economic conditions deteriorate. Bad economies inevitably create *legitimate reasons* to protest against oppressive and incompetent politicians. But eliminating transaction privacy enables politicians to predict *and instantly suppress* any form of civil disobedience *before it starts,* which prevents citizens from holding politicians accountable for their policies and actions.

Block Explorers Are Not Necessary to Keep Stakeholders Honest. The "block explorer" concept is one way to encourage honesty on a cryptocurrency network, *but it's not the only way.* Gini's BlockGrid architecture is designed to protect privacy *and* protect the integrity of the Gini ecosystem. Remember: "If you have nothing to hide, you have nothing to fear" was the infamous quote by the Nazi Propaganda Minister, Joseph Goebbels. Anybody who says they don't care about privacy because they have nothing to hide has no clue how essential transaction privacy is to democracy, political liberty and *all* human rights.[270]

Now back to the relatively technical stuff. . . .

The Global Transaction Vault. Gini transactions are private by default, but *certain* stakeholders may have a legitimate reason to verify *certain* transactions that have been executed by *certain* parties. So, the Gini BlockGrid architecture accommodates several transaction verification scenarios. It accomplishes this with Gini's decentralized Global Transaction Vault (GTV). The GTV is a logically separate decentralized ledger that contains all Gini transaction details. From an end-user's perspective, the GTV works like a very restrictive block explorer, but none of the transactions in the GTV are visible without the account owner's permission, just like in real-world auditing scenarios.

---

[270] Based on community feedback, we expect Gini's privacy features to evolve over time. For the latest updates on Gini's privacy features, visit: GiniFoundation.org/kb/privacy

Artificial Intelligence

A.I. Will Soon Dominate Every Aspect of Human Existence. In *Broken Capitalism: This Is How We Fix It*, I devoted an entire chapter to the consequences of A.I. on economic and political systems, but here is a brief summary:

The largest corporations and governments on Earth have irresistible economic and geopolitical incentives to use A.I. to manipulate *all* markets, exploit and control their customers and citizens, control the flow of capital and wealth in every economy, and generally render humans *and human liberty* obsolete. With the terrifying convergence of A.I., the IoT, robotics, and the alignment of corporate and government interests, gigantic corporations and governments will be unstoppable if we, as a species and as a human community, do not protect ourselves. The only way to protect ourselves is to build our own A.I. and deploy it like a guardian around every mission-critical system in our lives, including all Gini systems. Thus, it's important to clarify Gini's position on A.I. and how we will use A.I. to protect the integrity and stability of the Gini ecosystem.

A.I. is Already Destructive to Financial Markets & Broad-Based Wealth. Every major financial market has already been adversely impacted by automated trading robots for decades, which have sucked USD billions from the global economy.[271] However, over the past several years, more powerful A.I. has been controlling markets in unprecedented ways. "Nearly all market-making is presently dominated by machines that employ AI techniques. . . ." said Anthony Amicangioli, Wall Street Reporter for TheStreet.com.[272] In fact, for over two decades, A.I. has been evolving destructively in financial markets and it has primarily served to enrich a tiny number of entities who already have

---

271 CNN, B. L. van D., Hedge fund manager, Special to. (n.d.). High Frequency Trading turns exchanges turned into casinos. CNN.com/2014/04/07/business/opinion-van-dam-high-frequency-trading/index.html

272 Amicangioli, A. (2016, September 29). Artificial intelligence has rising impact on financial markets. Thestreet.com/story/13744043/1/artificial-intelligence-has-rising-impact-on-financial-markets.html

tremendous financial wealth and political power.[273,274] I have been pointing this out for many years; so, it's nice to see some other people on Wall Street finally admitting this is true.[275]

Free *Human* Markets Cannot Exist When A.I. Controls Markets.[276] Puritanical Libertarianism states that the solution to every problem is simply to *let the free market work,* but there is nothing *free* about a market that is overwhelmingly dominated by robots zapping your wealth away at nanosecond speeds. The Gini Foundation is founded upon the principle of "free *and* sustainable *human* markets," which is unattainable in a world dominated by A.I. and systemically entrenched gigantic corporations that do not have humanity's best interests at heart.

Existing Crypto Teams Have Too Many Conflicts of Interest. Every cryptocurrency project is launched based on a particular monetary policy and distribution of crypto wealth. The monetary policy and wealth distribution define and shape the structural economic, political, and cultural incentives of the project throughout its entire lifecycle. These structural incentives are more powerful than any other force because they determine which technical and economic options are *acceptable to the power players* at every step of the project's development. This is why the most influential humans (miners and crypto-rich whales) that dominate the major cryptocurrency projects today will never agree to reign in their own power to manipulate their markets. Thus, they will never be able to achieve sufficient consensus to fix the obvious A.I.- and monetary system problems that are already hurting the integrity and viability of their ecosystems.

How Can We Prevent Malicious A.I. from Damaging the Gini Ecosystem? The Gini Foundation will not hesitate to implement rational, market-stabilizing and ecosystem stability mechanisms based on

---

273 Biais, B., Woolley, P., & London School of Economics. (2011). High frequency trading. Manuscript, Toulouse University, IDEI.

274 Zagha, R., & Weltbank (Eds.). (2005). Economic growth in the 1990s: learning from a decade of reform.

275 Conversation, T. (2016, January 31). The real problem with high-frequency trading. Businessinsider.com/the-real-problem-with-high-frequency-trading-2016-1

276 Zhitnitsky, V. G. and I. (2014, April 3). Four Ways High-Frequency Trading Harms Investors & the Economy. Thestreet.com/story/12616225/1/fours-ways-high-frequency-trading-harms-investors-and-the-economy.html

several techniques, including:

- **Decentralized Gini Exchange.** Gini's decentralized exchange eliminates the need to depend on centralized gatekeepers that can lose or expropriate their customers' crypto and fiat wealth. This means the Gini exchange enables stakeholders to trade their assets directly with one another on a peer-to-peer basis with no corporate middle-men. Additionally, as a nonprofit entity, all Gini transaction fees are minimized and only intended to prevent market abuse and to ensure the Gini ecosystem is sustainable.

- **Speed Throttling.** The decentralized Gini Exchange includes automatic trade execution speed-throttling to prevent high-frequency trading (HFT) robots from manipulating the Gini cryptocurrency market in ways that are detrimental to real humans.

- **Systemic Bias *in Favor* of Humans.** A.I. already has a massive edge over humans; thus, we are not ashamed to declare that we intend to support humans in every structural way possible, including developing mechanisms to give higher weighting to every human transaction within the Gini ecosystem whenever our systems can accurately distinguish between humans and robots. For example, the decentralized Gini Exchange will perform technical behavioral analysis to distinguish between humans and robots; then it will automatically block aggressive robots that do not add any meaningful value to the Gini market. "Meaningful value" will be measured based on publicly published criteria to ensure transparency and integrity within the Gini ecosystem.

- **Meaningful Measure of Liquidity.** As discussed in the Gini article, "Robotic Market Volume is Not Market Liquidity," *liquidity* provided by robots today is often fickle and intended to trick human traders into buying/selling at the wrong time.[277] We

---

277 See the article "Robotic Market Volume is Not Market Liquidity" here:
GiniFoundation.org/kb/robotic-market-volume-not-market-liquidity

will stop this abusive tactic by measuring liquidity based on multiple criteria, including the length of time that a trader's buy/sell orders persist in the decentralized Gini market order book. These metrics will give the Gini Exchange protocol an automatic weighting mechanism so that high-quality liquidity is pushed to the top of the order book; while fickle and abusive liquidity is down-weighted.

- **Abusive Traders Will be Automatically Pushed Out of the Ecosystem.** In cases where A.I. or human traders abuse the Gini market, the Gini Exchange Protocol will automatically impose gradually increasing transaction fees on those offending traders until it is no longer profitable for them to continue their abusive behavior. As always, all these fees will be recycled back into the Gini ecosystem via the Value Streams System.

Clarity of Purpose Leads to Clarity of Action. In addition to the features above, Gini will not hesitate to implement mechanisms that a voting majority of the Gini Community believes is necessary to preserve the integrity and stability of the Gini ecosystem. Gini's highest priority is to create, preserve, and protect the integrity of the Gini ecosystem *for actual humans to engage in real-world commerce.* Thus, our purpose is to ensure that the Gini ecosystem is free from manipulation—*even if that means implementing mechanisms to block A.I. robots and manipulative traders*—because that's what will make the Gini ecosystem more stable and free than any fiat or cryptocurrency market on Earth today.

Gini Decentralized Exchange

The biggest problems that cryptocurrencies *are supposed* to fix are:

(1) the highly centralized fiat money supply;
(2) the highly concentrated economic power that plagues the fiat banking system;
(3) the corruption of the monetary system that leads to dysfunctional economies and political oppression.

One cryptocurrency team after another since 2008 has claimed that their cryptocurrency would liberate humanity from the tyranny of gigantic, monopolistic banks and oppressive governments. Yet, 10 years later, all the cryptocurrency markets *are even more centralized and concentrated* than the fiat banking system.

CRYPTO OLIGARCHY
7 ENTITIES CONTROL 91% OF BITCOIN NETWORK

BITCOIN CRYPTO OLIGARCHY IS SIMILAR TO MOST OTHER MAJOR CRYPTO-CURRENCIES TODAY.

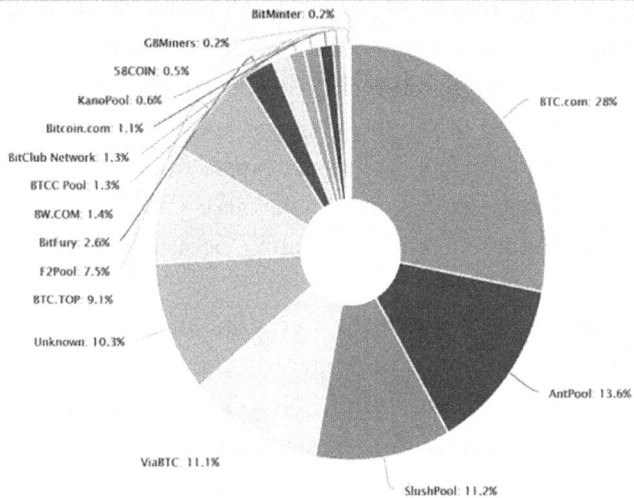

BitMinter: 0.2%
GBMiners: 0.2%
58COIN: 0.5%
KanoPool: 0.6%
Bitcoin.com: 1.1%
BitClub Network: 1.3%
BTCC Pool: 1.3%
BW.COM: 1.4%
BitFury: 2.6%
F2Pool: 7.5%
BTC.TOP: 9.1%
Unknown: 10.3%
ViaBTC: 11.1%
SlushPool: 11.2%
AntPool: 13.6%
BTC.com: 28%

Cryptocurrency Exchanges Are Highly Centralized. Today, the entire cryptocurrency universe revolves around a tiny handful of cryptocurrency exchanges. These crypto exchanges are highly centralized, *for-profit* corporations, which have become extremely powerful monopolies. In the U.S., *there are only two substantial exchanges* that enable users to convert fiat to cryptocurrency. (The same situation exists in most other countries.) This essentially positions these for-profit corporations as *central banks* at the center of the entire cryptocurrency universe.

Cryptocurrency Exchanges Are Now King-Makers. These centralized exchange corporations now have the power to pick and choose winners in the crypto markets. Specifically, they can arbitrarily determine which cryptocurrencies will be available to the general public;

and if you don't pay them a large sum of money (typically between $100,000 to $1 million) to list your cryptocurrency with them, they will simply ignore you.

Governments Have Total Control Over Centralized Cryptocurrency Exchanges. As for-profit corporations with enormous control over *other people's money* (just like gigantic fiat banks), these corporate cryptocurrency exchanges are now highly regulated by several governments. Currently, humanity's only relatively safe and cost-effective option to participate in cryptocurrency markets is to give our money *and trust* to highly centralized, for-profit corporate exchanges whose owners often conceal their identities and hide in the shadows. So, of course some government regulation is good in this context because it prevents some of the most egregious crimes that have been perpetrated by corporate cryptocurrency exchanges in the past.

However, the regulation of these exchanges has given politicians total power over them. On multiple occasions, governments have forced exchanges to indiscriminately hand over tens of thousands of private user accounts to potentially corrupt politicians and bureaucrats who may or may not have your best interests in mind.[278]

It's Difficult for Non-Technical Humans to Buy Cryptocurrencies. Ten years after Bitcoin's birth, it's still nearly impossible for the average, non-technical human to buy cryptocurrencies beyond the options offered by a tiny handful of centralized cryptocurrency exchanges. Why? Because these exchanges force people to go through a bunch of tedious, privacy-destroying steps just to open an account. Then the accounts are restricted to a tiny number of options. And overall, the exchange user interfaces are not as user-friendly as they could be. These factors are a significant barrier to wide-spread cryptocurrency adoption.

We Have Traded One Tyrant for Another Tyrant. Our nonprofit AngelPay payment processing company serves many thousands of for-profit merchants today; so, we certainly understand and appreciate people who want to make money. However, when

---

278 For many examples of institutional corruption, see "Is Institutional Corruption Only a Conspiracy Theory?" at: Eanfar.org/is-institutional-corruption-a-conspiracy-theory

central banks, quasi-central banks, and corporate monopolies position themselves at the center of all economic activity throughout an entire economy and use their power to extract profits for profit-obsessed shareholders and arbitrarily pick winners and losers in the marketplace, *that is real economic oppression.* Anybody participating in the cryptocurrency markets today has simply traded one tyrant (government-controlled central banks) for another tyrant (corporate-controlled central banks). Thus, humanity is no closer to true economic freedom today than we were when Bitcoin was first released in 2008 during the 2008 financial crisis, *which was caused by highly centralized and concentrated economic power.*

Centralized Cryptocurrency Exchanges Do Not Operate for the Public Interest. These for-profit exchange corporations have a strong economic incentive to encourage high-volume and automated speculative trading because they receive transaction fees for every trade that occurs on their exchanges *and* they seek to convert as many fees as possible into private profits. This means they have no meaningful incentive whatsoever to impose any meaningful rules that would reduce all the market manipulation that occurs on their exchanges. Moreover, like virtually all other cryptocurrency teams today, they hide behind the ideological façade of laissez-faire economics and Libertarianism to avoid taking any substantive action that would reduce the market manipulation and create more stability in any of the cryptocurrency markets.

Rampant Market Manipulation. We saw the Buy/Sell Walls in the previous chapter, which illustrate how a single whale can dominate and disrupt the entire market based on the way centralized exchanges work today.

What's the Real Purpose of Cryptocurrency Markets Today? We can clearly see that a single trader is able to completely dominate the entire market; and the same thing happens in all cryptocurrency markets today. These crypto whales, professional market manipulators, and superhuman A.I. bots can employ all kinds of predatory tactics (e.g., bear raids, wash trades, spoofing, pump-and-dump, front-running inside the centralized exchanges, and many more) to extract money from unsuspecting traders and destabilize markets whenever they want. This creates massive volatility, which makes it impossible to use any of the existing cryptocurrencies in real-world commerce. What's the point of all of this? To make a few crypto oligarchs and centralized corporate exchanges filthy rich and to concentrate millions of user accounts into a tiny number of choke points for politicians to control.

These Are All Philosophical Governance Problems, *Not* **Technical Problems.** These problems are not caused by any technical or legal barriers; it's a philosophical barrier. The teams behind these cryptocurrencies and exchanges have broken incentives and ideological blindness, which makes it impossible for them to ever fix any of these significant problems.[279]

Humanity is Still Waiting for Solutions. While existing cryptocurrency markets are mutating into new and even more powerful systems of economic oppression and political tyranny, the rest of the planet continues to wait for a cryptocurrency team that is serious, trustworthy, willing and able to fix all these problems. That's why Gini was born.

What Does Economic Freedom Look Like?

If humanity had a reliable, safe, *decentralized, peer-to-peer* currency exchange that enabled market participants to easily buy/sell crypto *and* fiat currencies directly with one another without any middlemen, then there would be no need for the cryptocurrency universe to be dominated by any centralized, for-profit exchanges at all. There would be no need for any potentially corrupt government regulation in the cryptocurrency markets because there would be no centralized entities to regulate. There would be no need for centralized corporations or governments to demand your ID and passport simply to exchange value directly with other humans. There would be no need or legal justification for governments to interfere with the direct, peer-to-peer exchange of value between humans. *That's* what true economic freedom looks like.

A Decentralized, Peer-to-Peer Exchange is Resistant to Tyranny. All democratic countries have some kind of constitution that protects an individual citizen's right to engage in private transactions with other individuals. Politicians in those countries would likely never

---

279 See the Gini School of Economics for a deeper analysis of these philosophical issues.

attack their citizens' right to engage in direct peer-to-peer trade because that would cross the line into the most extreme form of economic and political tyranny. That kind of extreme tyranny would have many far-reaching and long-term political and economic consequences inside their countries. No politician within a democratic country would be able to endure that kind of political pressure for long.

Gini's Decentralized, Peer-to-Peer Market Exchange. For all the reasons above, Gini is building a nonprofit, fully decentralized, *peer-to-peer* cryptocurrency exchange. This works like peer-to-peer file-sharing, but instead of exchanging files, Gini stakeholders exchange currencies, commodities and other assets. The Gini Exchange is designed to be very user-friendly. A screenshot of the Gini Exchange prototype is below.

The Gini Exchange Has Many Important Advantages. The Gini Exchange is designed to resolve all the significant problems that plague all the centralized cryptocurrency exchanges today. Below is a partial list of features:

- No central authority can steal or lose anybody's money.
- A.I. bots and professional market manipulators cannot manipulate the Gini Exchange market because transactions are executed directly with other peers on the exchange at the time of their choosing.
- The Gini Exchange is designed to be a safe and easy way to exchange fiat or other cryptocurrencies directly with other humans.[280]
- Crypto-to-crypto transactions are completely anonymous *and* private. No central authority (not even the Gini Foundation) can peek into your Gini Exchange account.
- The timing of the release of crypto or fiat funds to the transacting parties is controlled by the parties at all times.
- All transactions are protected by automated smart contracts and/or cryptographically secure escrow features, which prevents fraudulent trades.
- No human intervention from the Gini Foundation occurs unless one or both of the transacting peers requests direct arbitration of a dispute.
- It's unnecessary for governments to regulate the Gini Exchange because there's no central authority controlling other people's money.
- It's technically impossible for governments to regulate the Gini Exchange because there's no entity to regulate; the transacting peers do not transact through any centralized server or company. The Gini Exchange software connects them directly to each other just like peer-to-peer file sharing software.
- Individual users always retain complete control of their own currencies in their own accounts.
- The Gini Foundation does not charge any exchange fees for Gini currency transactions; so, our incentives are aligned with

---

280 See also: "Robotic Market Volume is Not Market Liquidity" at:
GiniFoundation.org/kb/robotic-market-volume-not-market-liquidity

the best interests of all Gini Exchange market participants.[281]

- The Gini Exchange will include the option to purchase precious metals (gold and silver) directly from small- to medium-sized mining companies in the developing world without any middlemen. This is a huge benefit for the poor communities where these companies operate and it's a great way for buyers to obtain precious metals more affordably than they could find at retail stores.

The Gini Exchange is an important component of the Gini Platform. We hope the partial list of features above gives humanity hope that there truly is a realistic path forward to fix the problems that plague the cryptocurrency markets today.

Simplify Your Life

We know that all the complexity in the cryptocurrency industry can be overwhelming for many readers. The easiest approach for evaluating whether to support a particular cryptocurrency project or not is to ask yourself the following questions:

- **Is their monetary policy sustainable?** By now, the reasons for this should be clear.
- **Do they have deep knowledge and real-world experience?** This includes experience in all the relevant technical domains, including online payment systems, cryptographic systems, network security, blockchain development, Political Economy, and ecommerce. Experience helps teams to avoid many mistakes and to avoid being distracted by temporary fads and gimmicks that waste time and resources.
- **Are their philosophical priorities aligned with your own priorities?** This is the most important factor because

---

281 Note: There's a difference between exchange fees and transaction fees. Virtually all blockchains have transaction fees because no blockchain is sustainable without some kind of transaction fee. Please see this page for more details: GiniFoundation.org/kb/gini-transaction-fees

technology is irrelevant if the team is using it to accomplish goals that are not aligned with your philosophical priorities.

- **Do they have a long-term vision and technical road map that is consistent with their stated philosophical priorities?** Many teams regurgitate fuzzy philosophical platitudes to add a veneer of purpose to their projects, but do they really have a unique and inspiring vision and a technical road map that is truly aligned with that vision?

- **Are they focused primarily on making money or making an impact?** There's nothing wrong with being focused on making money if the team is building a niche application for a particular industry or customer segment. However, if they're building *an actual cryptocurrency*, then they cannot be focused primarily on making money because they will be constantly tempted to compromise their ethics and make decisions based on profit motives rather than humanitarian motives. Most people understand this instinctively; so, they won't trust a cryptocurrency project that appears to be focused primarily on making money. Thus, over the long-run, projects that pretend to be focused on humanitarian goals while secretly trying to get rich will inevitably turn off many stakeholders, which means their projects will never be sustainable.

- **Does their project road map seem financially sustainable?** Vision, philosophy, great ideas and technology are all important, but does the team have a clear grasp of the budgeting and resource management requirements to be successful? Do they have experience building real companies or are they led by recent college graduates, lawyers, and investment bankers that have never really experienced the pain and joy of managing real R&D budgets, real people, and real companies?

When You Have Answers, You Have the Power to Take Action. After you ask and answer those questions for yourself, you should have a pretty clear idea of whether a particular project is worth your time and resources to support. Every project starts with nothing and evolves into something. So, the current state of a project at any particular moment is less important than feeling comfortable with the answers to the questions above. If you're comfortable with those answers, then getting involved and supporting an exciting cryptocurrency project throughout its evolution can be one of the most intellectually stimulating and gratifying experiences.

We hope our readers appreciate Gini's practical and humanitarian approach to building a cryptocurrency that achieves our primary goals: Protect human rights, provide an equitable and sustainable monetary system, and maximize the broad, wealth-generating potential of real-world commerce. If those goals are important to you, please consider visiting us at GiniFoundation.org and supporting Gini any way that you can.[282]

Key Points

- **A.I. Changes Everything.** Virtually nothing in human existence will be the same in the coming years. The convergent interests of politicians and gigantic corporations are already resulting in significant manipulation of political elections, financial markets and social networks. This is just the beginning. As A.I. becomes more powerful and embedded in every aspect of human existence, it will be virtually impossible for humans to detect when they are being manipulated. Gini is committed to developing A.I.-powered systems to protect humanity from malicious and exploitative A.I.

- **Ecosystem Stability is Gini's Highest Technical Priority.** Gini is committed to implementing rational technical mechanisms to preserve the stability of the Gini ecosystem to create the highest quality of life for all humans.

- **All Systems Require Trust.** Even when a system is fully automated, humans must still trust that the architecture and algorithms of the system are engineered according to trustworthy principles. Source code audits are useful to avoid

---

282 By the time you read this, the Gini Platform will have evolved in substantial ways. So, please visit GiniFoundation.org to see the latest whitepaper and articles about the Gini technology.

relying on trust *exclusively*, but no audit can truly capture all the implicit dependencies, assumptions and elements of trust that are inherent in all systems. That's why no other cryptocurrency team on Earth today has done more to earn the trust of humanity than the Gini team. Gini's systems have a built-in *continuum of trustworthiness* that ensures stakeholder privileges within the Gini ecosystem are always earned and incentives are always designed to achieve ecosystem stability, equity and sustainability.

- Chapter 9 -
The Gini Community

"Peaceful communities cannot exist without
sustainable and equitable economies." – Ferris Eanfar[283]

Empathy is the Foundation of Every Community. The capacity to understand and respond meaningfully to another creature's suffering is not unique to humans, but it's essential for human civilization. In fact, human civilization cannot exist without empathy, but true empathy is not about charity. Charity is merely a band-aid that masks the festering systemic injustices of broken economic and political systems that perpetuate poverty and modern slavery.

The Illusion of Progress. When the mass indignity of mass charity becomes the primary source of sustenance for the impoverished masses, we should not be surprised when human civilization degenerates into class warfare, ethnic conflict, and widespread violence. As the most evolved species on Earth, and with so many miraculous technological innovations at our fingertips, why does it often feel like humanity is speeding toward a dark, dystopian age?

Technology Alone is Not Enough. This is why the Gini team has spent many years analyzing the fundamental structure and real-world outcomes of many socioeconomic and geopolitical institutions, systems and events. In addition to the technologies we build, we've also written nonpartisan books and dozens of articles for our Gini School of Economics to share our research and observations about how wealth, power, economic, political and technological systems have evolved (or devolved) on Earth . . . and how they have become an inequitable and

---

283 I spent many hours trying to find another quote that captured the essence of this chapter. No existing quotes were adequate. So, please excuse this rare instance of self-quoting.

unsustainable noose that is rapidly squeezing the life out of capitalism and democracy on Earth today.

Most Cryptocurrencies Today Don't Solve any Meaningful Real-World Problems. Cryptocurrency teams often build technology for its own sake without making their tools and systems useful in real-world economic and governance systems. Or, they simply lack the skills, experience or authoritative access to the back-end banking and ecommerce infrastructure that's necessary to integrate their tools and systems with the credit card and banking systems that already exist. So, they build their tools and systems in ways that force their users to learn arcane protocols and user interfaces that are unfamiliar, risky, tedious and frustrating.

Gini is Different. The Gini team has proven, real-world experience building secure, scalable and user-friendly financial service systems, but we also understand that our technology is only as good as the *real-world* incentives, systemic integrity, and *systemic empathy* that we engineer into the Gini ecosystem. In fact, human civilization is only possible when empathy is engineered into the system. This is a fundamental principle in every technology that Gini builds, which is another reason why Gini is different from all other cryptocurrencies today.

The Law of the Jungle

To understand how and why the Gini ecosystem is based upon egalitarian principles and why broad-based wealth creation and distribution is so important to every human ecosystem, it's useful to briefly explore the biological basis of human cooperation.

Every Ecosystem is Governed by Something. Some ecosystems are governed by highly centralized governments, like in communist countries. Other ecosystems are governed by ruling elites who manipulate their economic and political systems to preserve their wealth and power, like in the United States and many other countries. Other ecosystems like Bitcoin and virtually all other cryptocurrencies today are governed by a tiny number of techno-oligarchs that have overwhelming crypto wealth and hash power within their networks. Finally, some ecosystems in nature are anarchic and have no premeditated governance structure at all; and thus, we often say they are governed by the *law of the jungle*.

Should Human Economies Be Ruled by the Law of the Jungle? Some people believe markets should never be touched by any kind of regulation. From their perspective, the *survival of the fittest* is the only way to structure an economy. They ignore the Capital-Labor Duality and all the cooperative elements and shared contributions that actually create value within every human economy and society.[284],[285] They ignore how adversarial competition is generally destructive and reduces overall value creation and capital formation. They often point to other animals in the animal kingdom and say short-sighted, shallow and self-serving things like:

All of creation is governed by the law of the jungle. Innovation, productivity, and all the things that make capitalism great are based on the survival of the fittest. Therefore, we cannot allow government to interfere in any way with the operations of a free market.

Free Markets are a Euphemism for the Law of the Jungle. We often hear people talk about *free markets* as if they were an economic reality, but pure "free markets" are a fantasy. In fact, pure free markets are simply another label for the "law of the jungle." With this awareness, it's easy to see why pure free markets are a fantasy: The law of the jungle inevitably leads to tyranny by the biggest and strongest entity in any ecosystem, which inevitably destroys every so-called *free market*.

Pure Free Markets = Tyranny. After a political or corporate tyranny is established, it's nearly impossible to overcome without a wealth- and capital-destroying violent revolution or international war. This is because the law of the jungle enables the biggest and strongest entities in the ecosystem to manipulate the political system and destroy anything that attempts to compete with their power. This destroys the so-called *free market* and leaves the victims of tyranny no choice but to

---

284 Nowak, M. A. (2012). Why We Help: The Evolution of Cooperation.
285 Axelrod, R., & Hamilton, W. D. (1981). The Evolution of Cooperation.

energetically protest and resist the tyranny, which often results in violent revolutions.

The Law of the Jungle Produces Monopolies. Puritanical libertarians believe government intervention is not necessary because gigantic corporations will eventually fail due to their own laziness, incompetent management, lack of innovation, etc. Of course, companies of any size can stumble and fall, but the phenomenon of perpetual returns to scale enables some companies to grow into gigantic corporate cannibals with so much inertia and power that it's technically and logistically impossible to escape their tyranny. At this point, they are a systemic risk to the ecosystem.

Long-Term Ecosystem Sustainability is a Higher Priority than Short-Term Self-Interest. John D. Rockefeller's Standard Oil company so thoroughly dominated the U.S. energy sector that no meaningful competition could ever emerge without government intervention. The same was true for AT&T, International Harvester, American Tobacco, and Microsoft. Other monopolistic companies today like Google, Amazon, Alibaba, Visa, Facebook, Walmart and others harm human ecosystems over the long-run, even though they claim to be giving consumers what they want in the short-run.

Obvious Problems Should be Easy to Fix, but Systemic Corruption Blocks Solutions. The idea that short-term self-interest is often destructive to long-term sustainability is obvious to any conscious adult human. So, why doesn't humanity already have more equitable and sustainable economic and political systems? Systemic corruption blocks thoughtful, rational humans from fixing these obvious problems. That means humanity will never see any meaningful solutions emerge from any major government, IMF, World Bank, World Economic Forum, or any entity that exists *within the global system* of economic and political control that was created by self-serving and short-sighted politicians and bankers in the U.S., Britain and France after World War II.[286]

The Human Animal. Because humans (especially certain humans with a genetic predisposition) have an endless appetite for wealth and

---

286 If you're not sure about this, read all the economic and geopolitical history books in the Gini Book List: GiniFoundation.org/kb/book-list

power, the human animal should not be perceived as merely another jungle citizen.[287] No, aggressive human animals are far more destructive: Instead of using their evolved intelligence to improve the quality of life for their species, they often use their intelligence to fabricate self-serving rationalizations about why they believe they're entitled to extract wealth from their species and the entire planet. Then, the self-destructive instincts in a relatively small segment of the human population are amplified by the behavior-corrupting incentive structures embedded within the toxic form of capitalism that dominates our planet today.

Supply & Demand in the Jungle. Human animals that subscribe to puritanical Libertarianism often use their intelligence to reduce the entire animal kingdom to the artificially simplistic construct of transactional supply and demand. But they conveniently ignore how the supply *and* demand for corruption and illicit wealth expropriation increases in a society when the supply of wealth and political power in the general population decreases. They ignore the fact that ecosystem stability and broad-based prosperity is a byproduct of broad wealth distribution, *not* concentrated wealth and power in the hands of a few jungle kings.

The Law of the Jungle Doesn't Work in Human Ecosystems. Humans share over 95% of their DNA and brain structures with other non-human primates. So, what separates humans from the rest of the animal kingdom? The computational power and plasticity of the human neocortex, which results in more complex culture. Cultural complexity produces a thin veneer of human civilization, which is an evolutionary adaptation that promotes sophisticated forms of cooperation. These unique features of human biology produce cultural resistance against any economic or political system that is based on the law of the jungle.

Human Biology Resists Pure Free Markets. The symbiotic biological human instincts for individual freedom and group cooperation are rooted in the human need for psychological certainty and ecosystem stability. Individual freedom gives us control over our own destiny, which maximizes certainty; human cooperation gives us

---

287 The genetic underpinnings of human aggression is an interesting topic. Learn more about that topic here: MyGeneFood.com/warrior-gene-5-common-myths

ecosystem stability, which maximizes the availability of food, water, shelter, companionship and other biological necessities. These instincts compel humans to fight against any form of tyranny that destroys certainty and ecosystem stability, which produces instinctual resistance against unsustainable concentrations of wealth and power.

Cooperation is Humanity's Equilibrium State. This might seem counter-intuitive at first, but the natural equilibrium state of every human ecosystem is a deeply cooperative environment that resists the self-destruction produced by anarchy and the law of the jungle. This equilibrium is only disrupted when a tiny number of sociopathic humans engorge themselves on wealth and power.

Cultural Vampires Create Cultural Violence. The unsustainable engorgement process inflicts trauma on the ecosystem, which predictably and inevitably triggers a violent, self-preservational response from other ecosystem inhabitants. This results in clan warfare, which, in the modern world, is amplified by profit-driven media companies and self-serving politicians who create, perpetuate and feed on the divisions within human societies like cultural vampires.

Debunking the Law of the Jungle Analogy. People who justify gluttony and greed with the "law of the jungle" argument conveniently ignore the fact that no other animal seeks to conquer and dominate all other species on the planet. Lions, bears, birds, elephants, fish . . . do not perpetually expand their dominion to consume all the resources and opportunities on the planet, thereby causing other groups of animals to starve and become corporate slaves to jungle kings. Virtually all other animals kill only what they can realistically eat. They never kill to hoard 1 million lifetimes worth of food and wealth in a planetary empire of gilded nests. Only short-sighted and gluttonous human animals do that.

Free Markets Are Necessary, but Not Sufficient. To be clear, a *relatively* free market that enables humans to engage in mutually beneficial value exchange is an essential feature of every stable and prosperous human society, which is why the Gini Decentralized Exchange exists. But free markets alone are not enough. The Capital-Labor Duality is the engine of value creation within every market-based economy and society. Thus, all fiat and crypto markets degenerate into the law of the jungle and corresponding destruction of broad-based wealth and political freedom whenever an ecosystem is dominated by a tiny number of gigantic corporate cannibals, gobbling up all the resources and opportunities within an ecosystem.

Being Rational, Self-Aware Humans is Not Activism. The Gini Community is not about *activism* per se; we are simply self-aware humans who are building technical systems and human ecosystems that are consistent with the laws of human nature, the laws of physics, and the laws of Mother Nature (i.e., Earth's ecological environment). The self-destructive fiat- and crypto-based economic systems that dominate our planet today violate every one of those areas of our planetary existence. Specifically:

- Perpetual corporate growth is not sustainable because the laws of physics constrain every system in the universe.
- Perpetual corporate economies of scale concentrates financial and political power in the hands of a tiny number of corporate executives/shareholders, which inevitably results in human rights violations, tyranny, and violent revolutions.
- Virtually all other animals kill only what they need to eat and they don't stomp on their ecosystems and strip the planet of all natural resources in a perpetual orgy of cannibalistic wealth-hoarding until there is nothing left for other animals.

No, we are not *activists*; we're simply rational humans doing what is necessary to create a more stable and sustainable planetary ecosystem.

Violent Revolutions

Violent revolutions are like volcanoes: The raw material that causes a sudden explosion of death and destruction is the predictable consequence of boiling elements beneath the surface. In 1788, King Louis XVI and his Ancien Régime enjoyed all the trappings of elite wealth and power. From within the velvet walls of the Palace of Versailles, the French royal family, nobility and clergy all believed their wealth and power were ordained by God. They believed the mere existence of their centuries-old regime *was proof* that the established order was the best system of economic and political governance ever conceived.

Within a few months, everything changed. By July of 1789, King Louis' kingdom was transformed into a frenzy of death and destruction. Tens of thousands of French elites were viciously murdered, sentenced to the guillotine without legal due process, their wealth was confiscated, and they were all violently removed from power. Throughout the entire process, all the way up to the moment of his death under the guillotine, King Louis was willfully blind to the reality of the real world: **Violent revolution *is inevitable* when economic and political conditions become intolerable to the masses.**

The same rapid disintegration of institutional order and violent confiscation of elite wealth and power has occurred many times in human history. Three times in a single human lifetime Russia was transformed *virtually overnight* from an economically oppressive czarist regime, to an economically oppressive communist regime, to an economically oppressive kleptocracy. Similar patterns of violence and tyranny have occurred in China, Chile, Germany, Argentina, Guatemala, Cuba, Korea, Italy, Iran, South Africa, Vietnam, Cambodia, and many other countries . . . *in the past 90 years alone.*

All Violent Revolutions Have a Common Theme. They occur because the economic pain and poverty of large human populations is substantially ignored by a ruling elite (aka, the Democide Class today). This happens whenever a political system structurally concentrates economic and political power into a relatively tiny number of hands. Just like King Louis and the elites of 18th-Century France, the fiat *and* crypto Democide Class in the U.S. and many other countries today are willfully blind to what is happening in the real-world outside the velvet walls of their unrealistic and self-serving economic theories and models.

In the United States and many other countries today, the raw material of violent revolution is boiling . . . and the temperature is rapidly rising. In fact, as predicted in my previous books and articles, violent protests are spreading throughout Europe and the Middle East today as I type. Our Gini team has accurately predicted the most significant socioeconomic events in recent years in the books and articles that we have published because we see clearly what is happening.

But *anybody* who understands economic history and the predictable cycle of violent revolutions should be able to see what is coming.

This is not about conspiracy theories, nor is this an attack on any particular special interest group, but let's be clear: The executives and largest shareholders of the banking cartel, gigantic transnational corporate cannibals, and self-serving Democide Class in many countries today are approximately 0.000001% of humanity, but they have created an economic and political system that is inequitable and unsustainable for the other 7.5 billion humans on Earth.[288] **By definition, anything that is unsustainable cannot last; thus, the status quo cannot and will not last.**

Economic Systems

When systems collapse, they must be replaced by something. However, economic system collapses create so much violence and pain that sociopathic dictators like Joseph Stalin, Adolph Hitler, and Mao Zedong *inevitably* emerge because only a sociopath can kill millions of humans to suppress the violence and chaos that ensues after an economic collapse. From a technical governance perspective, the problem with sociopathic dictators is that they are usually economic idiots, but the political power they obtain through violence gives them the delusion that they know how to solve their country's economic problems. So, they murder millions of humans because that's the only way to cram their unsustainable and idiotic economic experiments down the collective throat of their citizens.

Socioeconomic Operating Systems for Humanity. All economic systems exist along the same socioeconomic continuum. We know this is true because *highly centralized* governments, central banks, gigantic corporations and cryptocurrencies dominate global capitalism today; yet, there are substantial *market-based* structures in the communist regimes of China, Vietnam and the waning years of the Soviet Union. Of course, there is a clear *theoretical* difference between capitalism and communism, but anybody who claims there is a clear *real-world* distinction between

---

288 See the Gini Whitepaper for data to substantiate this: GiniFoundation.org/gini-whitepaper

"free-market capitalism" and "authoritarian communism" is mistaken. Socioeconomic systems manifest *in the real world* along a continuum and they can be *configured* to produce optimal outcomes just like a computer operating system.

Optimizing for Broad-Based Wealth & Sustainability. Nothing in macroeconomics happens by accident. Every fiat and crypto monetary system (including Gini's) is driven by an agenda. The agenda either optimizes for *broad-based* wealth and power or it optimizes for wealth and power concentrated in the hands of a tiny group of ruling elites. In a *democratic* society, politicians and cryptocurrency project teams *should be configuring their capitalism* according to their stakeholders' unique cultural and philosophical values and priorities, which would optimize their socioeconomic operating systems to maximize the wealth and quality of life for the largest number of humans. Unfortunately, that is not the intention of politicians, their corporate overlords, and other cryptocurrency projects in many countries today, which is why broken capitalism exists.

Capitalism Comes in Different Flavors. Some people believe capitalism is fundamentally unsustainable, but as discussed in the Gini article "Is Capitalism Sustainable?", the sustainability of capitalism is not a binary, yes/no question.[289] In fact, there are at least five distinct *flavors* of capitalism: agrarian capitalism, industrial capitalism, financial capitalism, bank capitalism, and Gini Capitalism. Capital, value, wealth, political power and quality of life flow throughout a society according to how politicians *configure* their country's socioeconomic operating system. Thus, "capitalism" is not a monolithic system that we can logically say is *good* or *bad*.

What is the "Capital" in Capitalism? Capitalism is how stored value and wealth ("capital") is organized and deployed to capital-intensive projects that produce value, income, wealth and human happiness within an economy. A society *can choose* how it wants to *configure* the value creation and distribution process within their economy. Thus, *capitalism can and should be configured based on the prevailing values and priorities that are important to a broad majority of humans in each*

---

289 See: GiniFoundation.org/kb/capitalism-sustainable

society, not based on the narrow interests of a tiny number of elites, crypto whales or shareholders of gigantic corporations. This is the only way to ensure broad-based peace and happiness, which is the only way to avoid violent revolutions that destroy capital, wealth, happiness, communities, cultures, and entire countries.

Capitalism Mutates as an Economy Grows. As an economy grows and becomes more diversified, it graduates from agrarian capitalism to industrial capitalism. As a country's industries produce more capital-intensive products, it graduates to financial capitalism, which is necessary to aggregate financial capital to build larger factories and infrastructure projects. Unfortunately, financial capitalism is relatively easy to centralize and control by manipulating and controlling the financial and industrial regulations that are produced by a country's political system. In the fiat world, this is why gigantic banks and their largest shareholders *inevitably* hijack governments and exploit them for their own private interests. When this happens, the economy has mutated into *bank capitalism*.

Bank Capitalism Destroys Capitalism. Bank capitalism enables banks to insert themselves like a deadly virus into the middle of all economic activity throughout every economy on Earth today.[290] Then, like a giant vampire-squid, gigantic banks are able to suck value out of the global economy and instigate wars that amplify their profits, wealth and power. This is how the global economy has mutated into the most self-destructive type of capitalism on Earth today: bank capitalism.

Trickle-Down Economics Does Not Work. Politicians and crypto teams can configure their socioeconomic operating systems to maximize broad-based wealth and power OR maximize corporate/crypto whale profit, but they cannot maximize both because maximizing corporate/crypto whale profit is mutually exclusive to maximizing broad-based wealth and power. Additionally, we know trickle-down economics does not work for many reasons, which is why a society must choose between maximizing broad-based wealth/power

---

290 Even the Bank for International Settlements admits this. To confirm this and all the problems associated with bank capitalism, see the Gini Whitepaper on the Gini website to view the empirical data and scientific third-party reports linked therein.

or maximizing corporate/crypto whale profit.[291] The choice we make determines our flavor of capitalism and how capital, value, wealth, political power, happiness and quality of life are distributed throughout a population. This truth is the basis of our name—"Gini"—which is based on the Gini Index, the gold standard in Economics for measuring the distribution of wealth within every country.[292]

Capitalism vs. Communism Propaganda. The fundamental humanitarian problems that the Gini Platform is designed to resolve have nothing to do with the philosophically shallow noise and propaganda associated with the *communism vs. capitalism* debate. Although many people believe in various forms of socialism, very few humans on Earth today actually believe economic communism works because the destructive outcomes of economic communism are easy to see for any human with a brain.

Regardless, Gini is a nonpartisan organization and communism has nothing to do with the Gini philosophy and technology. Anybody who falsely accuses Gini of any *–ism* obviously does not understand this basic reality: Violent revolutions are the *predictable and inevitable consequence* when wealth and political power are concentrated into the hands of a tiny group—the Democide Class—which creates a socioeconomic volcano that inevitably explodes into widespread death and destruction.

Gini Capitalism

Human civilization is propelled by incentives; and all human ecosystems are shaped by *incentive structures*, which are combinations of interdependent incentives embedded within every human environment. This is true in socioeconomic systems like capitalism and democracy and it's true for all human governance systems, including the political institutions of governments, profit-driven boardrooms of banks and corporations, cryptocurrency projects, and nonprofit humanitarian organizations. The Gini team has spent years analyzing and developing systems that have rational, equitable and sustainable incentive structures

---

291 More data regarding Trickle-Down Economics here: GiniFoundation.org/kb/trickle-down-economics

so that the Gini ecosystem is as equitable and sustainable as humanly possible.

Does Gini Have the Right Incentives? For all the reasons discussed in the Gini article "Who Has the Right Cryptocurrency Incentives?" the Gini Foundation is structured as a tier-1 nonprofit NGO to ensure that Gini always has sufficient financial and political independence to avoid being captured by politicians and special interest groups.[293] Based on the analysis in that article, it's difficult to find any other organization like Gini that has properly aligned incentives *and* the technical skills, knowledge and *proven* real-world experience to build a privacy-assured cryptocurrency that protects human rights, provides an equitable and sustainable monetary system, and maximizes the broad-based, wealth-generating potential of real-world commerce.

After recognizing all the real-world realities that we have discussed so far, the next logical questions are: What does a sustainable economic system look like? What alternative exists that can compete with the toxic and unsustainable status quo? Let's focus on that now.

Gini Capitalism is an Easy Way to Gradually Opt-Out of the Toxic Status Quo. Gini's Ecosystem Stability Mechanism (ESM) enforces money supply best practices by ensuring that the ecosystem's Gini Index never exceeds 25% for too long. This has many profoundly positive consequences for an economy because it automatically prevents corporate cannibals from rampaging all over the ecosystem and destroying the Capital-Labor Duality at the heart of *every type of capitalism*. The ESM and the other Gini systems discussed previously are the essence of a socioeconomic operating system for humanity that we call "Gini Capitalism," which is far more equitable, democratic, and sustainable than bank capitalism. Gini Capitalism is also an easy and fun way for humanity to gradually opt-out of the toxic status quo.

How Does Gini Capitalism Work? In contrast to self-destructive bank capitalism, Gini Capitalism uses technology for humanitarian purposes. Specifically, the Gini technology anonymously quantifies the value creation process wherever and whenever value is created within

---

292 To learn more about the Gini Index, see: GiniFoundation.org/kb/gini-index
293 See: GiniFoundation.org/kb/cryptocurrency-incentives

the Gini ecosystem. This enables Gini's Value Streams System to reward stakeholders according to their contributions to the ecosystem. This ensures that unearned power based on race, social class, political connections, and other non-value-creating factors in the broken fiat system do not skew the distribution of wealth and power in the Gini ecosystem.

Gini Capitalism is Inherently Stable. The ESM ensures that no single entity can ever become a systemic risk to the ecosystem. Then, Gini's Community Governance System ensures that all decision-making processes that significantly impact the ecosystem are based on the most egalitarian and democratic voting process humanly possible.[294] This ensures that no cartel can become a systemic risk to the ecosystem. Together, these mechanisms ensure that Gini Capitalism will always be the most equitable, stable, democratic and sustainable socioeconomic operating system humanly possible.

Gini Capitalism Can Save Humanity from Bank Capitalism. Gini Capitalism is a viable and credible alternative model for how a socioeconomic operating system for humanity can work. Of course, we don't expect Gini Capitalism to replace the toxic status quo overnight. We don't expect gigantic banks and corporations to gleefully embrace any of Gini's principles, but the Gini Platform is designed to gradually liberate humanity from cannibalistic banks over time.[295]

Nothing Will Change if We Don't Take Responsibility for Our Own Destiny. Gini Capitalism (or any meaningful reform) must grow from the grassroots in every community. Nothing will ever change if we wait for self-serving politicians and gigantic corporate cannibals to give us solutions because their incentive structures prevent them from ever fixing anything. We, the people, must take responsibility for our own destiny; and Gini Capitalism is a viable path forward.

"But Gini Capitalism Doesn't Include XYZ. . . ." We have many new features and ideas to share and explore with the Gini Community over time. Please remember: This is just the beginning; so, the Value Streams System and other Gini Platform features will likely

---

294 See the Community Governance System here: GiniFoundation.org/kb/community-governance-system

expand to include and reward other forms of value creation over time. The Gini technology is inherently designed at every level to amplify and reward the human value creation process and eliminate the toxic elements that prevent humanity from achieving its full economic and humanitarian potential today. This is the best foundation possible, but we, as a human community, will grow and improve upon this foundation over time.

Time is Running Out

When bank capitalism inevitably collapses, what will replace it? If we, as a human community, don't start working to replace it with something now, it will be too late after the collapse because tyrannical dictators will rise again and murder anybody who resists their idiotic economic experiments. That's not fear-mongering; it's the *predictable outcome* of the chaos and violence that occur after a major economic collapse.[296] If we don't act now, then we can expect somebody else with far less altruistic intentions will act later; and humanity will suffer a far worse fate under the iron grip of the next Stalin, Hitler or Mao.

We Can Prevent a Violent Systemic Collapse. Rising economic and geopolitical tensions today are already creating the conditions for global war, which are fundamentally caused by the unsustainable bank capitalism that is choking the life out of humanity today.[297] Gini's purpose is to prevent a violent systemic collapse by creating an alternative path forward. This path is a way for humanity to gradually opt-out of the toxic status quo. The path is powered by unique, community-driven open source technologies that are explicitly engineered to protect human rights, provide an equitable and sustainable monetary system, and maximize the broad-based, wealth-generating potential of real-world commerce.

---

295 Learn more about this here: GiniFoundation.org/gini-credit-union

296 NB: The 2008 financial crisis was *not* a collapse; it was just a crisis. The Great Depression was a collapse, which led to over 70 million human deaths. To learn more, see: Eanfar.org/birth-modern-welfare-state

297 See "Is U.S. Foreign Policy Controlled by Corporations?" for more context: Eanfar.org/is-u-s-foreign-policy-controlled-by-corporations

Authentic Alignment

Prioritizing Human Life and Happiness Over Profit. Many organizations seem to be guided by nothing more than the obligatory *customer-is-king* lip-service. Even when they start out on a good path, they often abandon their original core values because their profit-obsessed corporate structure and short-sighted culture are fundamentally engineered to squeeze every penny from their customers, which creates endless conflicts of interest. This is what causes them to behave in ways that destroy the Capital-Labor Duality. As a nonprofit organization, The Gini Foundation is structurally and culturally engineered to maximize human life and happiness—not profit—but we ensure that the value we deliver to the communities we serve is financially sustainable.

History of Innovation & Positive Impact. For decades, the Gini team has directly contributed to the evolution of the global financial services sector. When we invented and founded Authorize.Net in 1996, there was no secure and scalable way to process payments and credit cards over the Internet. The technologies we invented gave birth to the global ecommerce sector and continue to empower billions of humans worldwide, reduce poverty in virtually every country, and generate trillions of dollars of new wealth throughout the global economy. Today, Authorize.Net is the largest payment gateway in the world and has processed over $1 trillion to date.

The Genesis of Gini. After we sold Authorize.Net, we had some time to think deeply about the many problems that have plagued the financial services sector for decades. Our solution to some of these problems is the nonprofit AngelPay Foundation. Founded by the most experienced team in the global payments industry, AngelPay is one of the ways that we have given back to the merchant communities that have been good to us over the years. However, AngelPay is only focused on the payment processing industry. The problems described in our books and articles extend far beyond the payment processing industry, which is why Gini was born.

A Growing Community. The Gini team is using our skills, technologies, and resources to provide nonprofit services, tools and leadership to serve the public interest, not the interests of a tiny group of shareholders and ruling elites. Gini's nonprofit legal structure and strong focus on Gini's deeply meaningful humanitarian mission ensure that our incentives are always fundamentally aligned with the interests of the communities we serve. This enables us to authentically promote the common good and improve the quality of life in the communities and organizations that participate in the Gini ecosystem.

World-Class Team. The nonprofit services that Gini provides are delivered by the same team that pioneered secure and scalable ecommerce when we invented and founded Authorize.Net in 1996. Additionally, the Gini team includes former senior executives with experience in many other industries, professors from academia, and other like-minded business and community leaders. This proven and diverse *nonpartisan* experience has given us deep technical, geopolitical, and sociological insights into how to resolve the problems that plague the global economy today.

Gini's Interests Are Aligned with the Community. Gini is the first and only nonprofit cryptocurrency that is designed with an equitable and sustainable monetary system, serious transaction privacy, and focused on maximizing the broad-based, wealth-generating potential of real-world commerce. We're not aware of any other team that has the same combination of real-world experience, technical skill, philosophical awareness, financial independence, deep understanding of real-world economics, and structural incentives to operate the way Gini does. In contrast, the monetary systems and incentive structures of other cryptocurrency projects make it structurally impossible for them to operate without succumbing to many inherent conflicts of interest.

Proven Experience Overcoming Adversity. When we founded Authorize.Net in 1996, we literally started the company from a garage, scraping and clawing every penny to pay our bills. That experience was incredibly difficult in the beginning because very few people (including all the major banks) believed it was possible to build a secure and scalable system to accept payments and credit cards over the Internet. Fortunately, we persevered and eventually we built Authorize.Net into the most recognizable brand in the global online payment processing industry. Our experience building a truly global company from the ground up and serving tens of thousands of merchants and millions of customers worldwide has given us deep insight into how to grow Gini Capitalism into a truly viable alternative to the toxic status quo.

The End of a Book; the Beginning of a Journey

Significant Change Rarely Happens Overnight, but What Choice Does Humanity Really Have? Regardless of their intentions, the political creatures in governments around the world *will never* fix any of these problems because they are trapped in broken incentive structures. If we don't build a viable alternative to the toxic status quo now before the next great economic depression, another Hitler, Stalin or Mao with much less noble intentions will impose their short-sighted and tyrannical agenda on humanity. Therefore, we ask the non-political humans of Earth to stand together with Gini against tyranny. Together, with our strength in numbers, we can create a better socioeconomic operating system for humanity . . . before it's too late.

Thank You. The Gini team and I thank you for reading this book. By now, it should be clear that Gini is different from other cryptocurrency projects in nearly every conceivable way. Writing this book has been a gratifying journey, but the best is yet to come. Now, we invite you to visit GiniFoundation.org to participate in the growing Gini Community at this critical moment in human history. Together, we really can make a meaningful difference.

Key Points

- **Peaceful Communities Cannot Exist without Sustainable and Equitable Economies.** To avoid humanitarian catastrophes, human societies should do whatever is necessary to ensure that economic and governance systems produce sustainable and equitable outcomes.
- **Empathy Must be Engineered into the System.** Empathy is the foundation of human civilization. That's why it's not enough to build economic and political systems with "systemic integrity." "Systemic integrity" is meaningless if the definition of "integrity" does not include ecosystem stability and equitable

sustainability, which is how empathy manifests within institutional systems. No system has integrity if it can be manipulated and abused by a tiny number of oligarchs, whales and the Democide Class who skew the distribution of wealth and power toward themselves.

- **Violent Revolution is Inevitable when Economic and Political Conditions Become Intolerable.** All of the Gini team's socioeconomic predictions are coming true today. By definition, anything that is unsustainable cannot last; thus, the status quo cannot and will not last. When bank capitalism inevitably collapses, something must replace it. Gini Capitalism is a viable alternative to bank capitalism.

Before You Go . . .
Why Are Book Ratings Important?

In today's hyper-connected world, books live and die by online ratings. The algorithms at online bookstores push books up or down their book lists based on how many positive ratings they receive. Thus, without positive ratings from thoughtful readers like you, this book might not be seen by enough people to make the positive impact that it can make. If you appreciate this book, please share it on social media, follow Gini on Twitter (@GiniFoundation) and post a positive rating wherever you obtained it. Every positive rating and media share counts because they add up over time and help draw attention to Gini's important principles and solutions.

Have Critical Feedback? If you have critical feedback, please send it privately to GiniBook@GiniFoundation.org. All feedback is appreciated, but negative public feedback is not necessary when an author or publisher provides a feedback mechanism directly to readers. Even if you disagree with the style or content of this book, hopefully you can still appreciate Gini's important purpose. Sharing your critical feedback privately will help us improve Gini's weaknesses without slowing down Gini's community-building progress. Thank you.

About the Author

You can visit Ferris' LinkedIn page (LinkedIn.com/in/ferriseanfar) for the most up-to-date summary, but in brief: Ferris Eanfar has over 20 years of experience in technical, financial, and government intelligence environments. He is a Senior Partner at Vision Bankcard; co-founder of the nonprofit AngelPay Foundation, the first and only nonprofit financial services company dedicated to returning wealth and power to the creators of value; and co-founder of the nonprofit Gini Foundation, which builds unique cryptocurrency systems that protect human rights, among other important benefits. Ferris' professional background includes payment/credit card processing, asset management, commodities trading, artificial intelligence software engineering, blockchain/cryptocurrency development, and military and government affairs. He is a U.S. Air Force veteran and he worked in the U.S. Intelligence Community as a Cryptologist and Linguist with a Top Secret (TS/SCI) security clearance. He has written over 100 articles and several books in the fields of International Political Economy and blockchain/cryptocurrencies, including the "Global Governance Scorecard," the "Blockchain Cryptonian," *Broken Capitalism: This Is How We Fix It* and *GINI: Capitalism, Cryptocurrencies & the Battle for Human Rights.*

Ferris enjoys spending time with family and friends, traveling to interesting places, playing and composing piano music, and writing nonpartisan books and articles about the geopolitical, economic, social, technological, and philosophical phenomena that impact large populations. Ferris has numerous technical and financial certifications and his formal education was in Cryptological Linguistics at the Defense Language Institute and International Political Economy at Penn State University.

Index

www.ingramcontent.com/pod-product-compliance
Lightning Source LLC
Chambersburg PA
CBHW060352220326
41598CB00023B/2894